Mastering Optics

Mastering Optics

An Applications Guide to Optical Engineering

John Blackwell and Shane Thornton

McGRAW-HILL BOOK COMPANY

London · New York · St Louis · San Francisco · Auckland · Bogotá · Caracas
Lisbon · Madrid · Mexico · Milan · Montreal · New Delhi · Panama · Paris
San Juan · São Paulo · Singapore · Sydney · Tokyo · Toronto

Published by

McGRAW-HILL BOOK Company Europe
Shoppenhangers Road, Maidenhead, Berkshire SL6 2QL, England
Telephone 01628 23432
Fax 01628 770224

British Library Cataloguing in Publication Data

Blackwell, John
Mastering Optics: Applications Guide to
Optical Engineering
I. Title II. Thornton, Shane
621.3827

ISBN 0-07-707875-6

Library of Congress Cataloging-in-Publication Data

Blackwell, John (John Michael)
Mastering optics : applications guide to optical engineering/
John Blackwell and Shane Thornton.
p. cm.
Includes bibliographical references.
ISBN 0-07-707875-6 (pbk. : alk. paper)
1. Optics. I. Thornton, Shane (Shane William).
II. Title.
TA1520.B56 1996
621.36—dc20 95–47412
CIP

McGraw-Hill

*A Division of the **McGraw-Hill** Companies*

1234 CUP 9876

Typeset by Mackreth Media Services, Hemel Hempstead, Herts
Printed and bound in Great Britain at the University Press, Cambridge

Printed on permanent paper in accordance with ISO Standard 9706

Contents

Preface **ix**

Acknowledgements **x**

1 Basic optics **1**

Introduction 1

Refractive index 1

Refraction 2

Paraxial optics 6

Sign conventions 7

Lens shapes 8

Lens equations 9

Magnification 10

Real and virtual images 12

Principal planes and nodal points 12

Ray nomenclature 16

Lens nomenclature 17

f-number 18

Power 19

Resolution 19

Diffraction 20

Airy disc 20

Modulation transfer function 21

Depth of focus 23

Aberrations 25

Polarization 33

2 Lenses and coatings **35**

Introduction 35

The singlet lens 35

The doublet lens 38

	The triplet lens	40
	Complex lenses	42
	Microscopes	45
	Aspheric lenses	49
	Fresnel lenses	51
	Mirror lenses	52
	Coatings and filters	54
	Reflection at a material boundary	55
	Anti-reflection (AR) coatings	57
	Multilayer filters	59
	Dichroic coatings	60
	Hot and cold mirror coatings	63
	Gradient index (GRIN) coatings	64
	Specifying a coating	64
3	**Prisms and diffraction gratings**	**65**
	Introduction	65
	Prisms	65
	Types of prism	67
	Types of diffraction grating	73
	Diffractive optics	79
4	**Reflectors and mirrors**	**83**
	Introduction	83
	Optically flat mirrors	83
	Curved mirrors	86
	Materials	97
5	**Commercial components**	**100**
	Introduction	100
	Buyers' guides	100
	Mail-order lenses	101
	CCTV camera lenses	103
	35-mm camera lenses	104
	Microscope lenses	104
	Custom design	104
6	**Measuring light**	**106**
	Introduction	106

	Photometry	106
	Energy	107
	Power	107
	Luminous intensity	107
	Flux density	108
	Illuminance	109
	Cosine laws	109
	Illuminance and lenses	110
	Measuring illuminance	111
	Screen lumens	111
	Lighting design lumens	112
	Illuminance conversion factors	112
	Exitance	112
	Brightness	113
	Luminance conversion factors	114
	The human eye	114
	Rods and cones	114
	Photopic response	114
	Scotopic response	116
	Mesopic response	116
	Spectral luminance efficacy	117
	Sensing colour	118
	Colour measurement and chromaticity coordinates	118
	Chromaticity coordinates	119
	Uniform chromaticity diagrams	119
	Colour differences	121
	Detectors	121
	Detection by a single element	122
	CCD camera	132
	Detection systems and good housekeeping	133
7	**Fibre optics**	**138**
	Introduction	138
	Fundamentals	138
	Fibre types	141
	Connectors	145

	Gradient index (GRIN) lenses	149
	Fibre preparation	150
	Launching light into fibres	151
	Applications for fibres	152
	Incoherent bundles	152
	Coherent bundles	153
	Communications	153
	Fibre sensors	153
	Mode stripping	157
8	**Light sources**	**158**
	Introduction	158
	Lasers	158
	Laser types	169
	Laser applications	171
	Light-emitting diodes	175
	Edge-light-emitting diodes	180
	The laser diode	180
	Fluorescent sources	184
	Blackbody radiation	185
	Tungsten halogen lamps	186
	Spectral radiant exitance	187
	Discharge lamps	189
	Short arc lamps	190
	Long arc lamps	191
	Spectral lamps	192
9	**Mechanics**	**193**
	Introduction	193
	Lens drawing and specifications	193
	Lens mounting	204
	Component mounting: fibres and prisms	206
	Good practice	209
Appendix A	**Glossary**	**213**
Appendix B	**List of symbols**	**234**
Appendix C	**Mathematical formulae**	**236**
Further reading		**245**
Index		**246**

Preface

We have attempted to write an informative text at an exact but not academic level to impart a basic understanding of modern optics. It is aimed particularly at engineers, working in optical engineering for the first time, who come from other disciplines such as mechanical or electronics engineering.

There are two major sections: the body of the text and the appendices. The latter have been included so that unfamiliar terms can be quickly understood and common formulae rapidly found.

Where an in-depth treatise is required, the list of further reading material presents a number of books, all of them well known to the authors: they come with the highest commendations. But, however much you read there is no substitute for getting out there and doing it!

Acknowledgements

Throughout the course of writing this book we have benefited greatly from the constructive comments of our reviewers and colleagues. We have also appreciated the support and patience of our editors at McGraw-Hill.

CHAPTER 1

Basic optics

Introduction

This first chapter describes the basic terms and formulae that are commonly used in the design of optical systems. We begin with the most fundamental concepts in optics, namely, refractive index, refraction and Snell's law, then move on to discuss lenses, depth of focus and resolution. This leads naturally into a description of the different types of aberrations and their relevance to optical systems.

No attempt is made to derive in detail the formulae that appear. This is a deliberate decision based on the experience that in industrial engineering the result is more important than the intricacy of the mathematics. Consequently, we have concentrated on the relevance of this basic optics knowledge to real-life examples, and on simplifying each formula to an easily used form. However, anyone requiring a greater mathematical insight should refer to Appendix C.

Refractive index

The refractive index, n, of a material is the ratio which relates the velocity of light in that material, c_m, to the velocity of light in a vacuum, c_v. For most common applications the light is always slowed down by its passage through any medium other than a vacuum, and so the ratio is almost always greater than 1.0.

$$n = c_v/c_m \tag{1.1}$$

For most practical purposes the velocity of light in air is used rather than in vacuum. To be exact, the refractive index of air at 15 °C is 1.0003 but it is assumed to be 1.0 for most calculations. The refractive indices for some common optical materials in green light at 546 nm are set out in Table 1.1.

Table 1.1 Refractive indices of some common materials

Material	*n*
Optical crown glass (BK7)	1.5187
Pyrex	1.4727
Optical flint glass (SFL6)	1.8126
Sapphire	1.7708
Acrylic	1.4938
Water	1.33

The velocity of light in a material, and hence the material's refractive index, depends on the wavelength of the light. There are a number of different formulae for calculating the refractive index at a given wavelength, but the one most commonly used by glass manufacturers is

$$n^2 = A_0 + A_1\lambda^2 + A_2\lambda^{-2} + A_3\lambda^{-4} + A_4\lambda^{-6} + A_5\lambda^{-8} \tag{1.2}$$

where λ is the wavelength in micrometres and A_0 to A_5 are constants for the material. Glass manufacturers usually give all six coefficients for each glass, and the formula is accurate to better than 1 part in 10^{-5} from 350 nm to 1 μm.

Refraction

A wavefront of light moving at an angle across the boundary between two different materials will be deviated, or refracted, at the interface. If the wavefront is moving from a low to a high refractive index then refraction always decreases the angle that it makes with a normal to the boundary surface. This is illustrated in Figure 1.1, which shows how a plane wavefront is gradually slowed down and deviated as it enters a flat piece of glass.

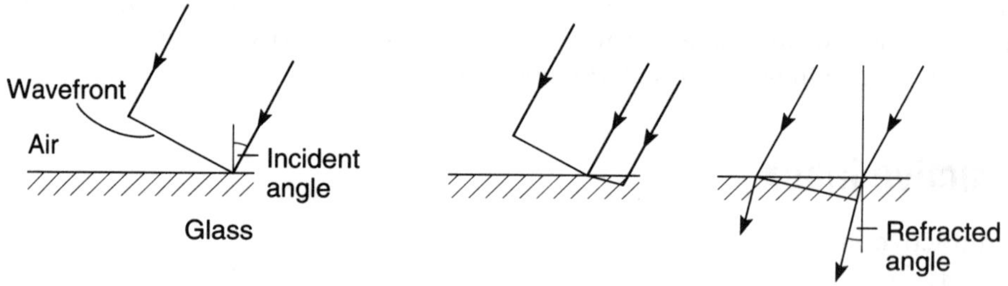

Figure 1.1 Refracted wavefront

Refraction can be a sudden change in direction caused by a clearly defined index boundary, as with an air–glass interface. Or it can be a gentle change extended over some distance as the ray traverses a region of slowly varying index. Examples of this latter type of refraction occur in, among other places, a light beam confined by a graded index multimode, optical fibre, and a mirage, where local heating of the air close to a hot surface has caused a decrease in the refractive index of the air in that area.

Snell's law

Refraction and refractive index boundaries lead naturally to Snell's law, which is the formula that allows us to calculate the angle of refraction at a refractive index boundary. Angles are always calculated with respect to a normal to the boundary surface. For an angle of incidence θ_i in a material with an index n_1, the angle of refraction θ_r in a material n_2 can be calculated from

$$n_1 \sin\theta_i = n_2 \sin\theta_r \tag{1.3a}$$

or, for light passing from air to another material n_2, from

$$\sin\theta_i = n_2 \sin\theta_r \qquad (1.3b)$$

This two-dimensional form of Snell's law forms the basis for all ray tracing and can be applied as follows.

Figure 1.2 represents an air-to-glass boundary with a ray making an angle of 30° to the surface normal. Assume that the glass type is BK7 (see Table 1.1) and that we are working with light of wavelength 546 nm, which is the green mercury 'e' line. We know from manufacturers' data that the refractive index of BK7 at this wavelength is $n_e = 1.5187$. Rearranging Snell's law gives

$$\theta_r = \sin^{-1}[(n_1 \sin\theta_i)/n_2]$$

i.e. a refraction angle, θ_r, of 19.2°.

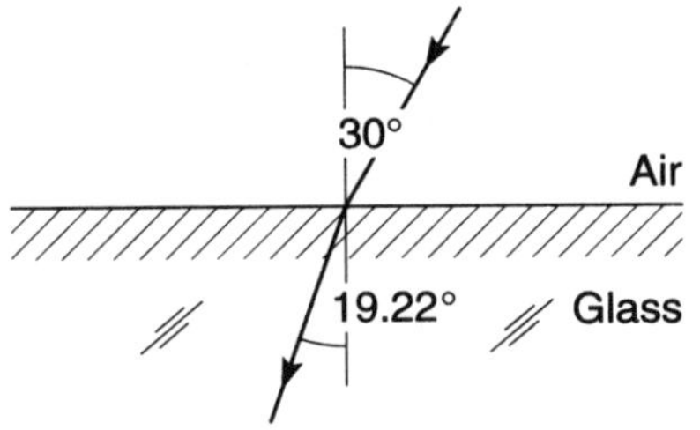

Figure 1.2 Refraction at a plane surface

If we now consider the air-glass interface as part of a plane, parallel-sided glass block we can calculate the magnitude of refraction as the ray emerges from the block. Under these circumstances, and following the same nomenclature as before, we have $n_1 = 1.5187$, $n_2 = 1.0$, $\theta_i = 19.2°$.

Snell's law gives:

$$\theta_r = \sin^{-1}[1.5187 \times \sin(19.2)/1.0003]$$
$$= 30.0°$$

This result is illustrated in Figure 1.3 and shows that for a ray traversing a plane, parallel-sided glass block the input and exit angles are identical. This applies as long as the refractive index of the block is homogeneous.

An equally important result from this example is that the optical path of the ray has been reduced, in other words the 'apparent' or optical thickness of the block is less than its mechanical dimension. To remember this, think of looking into a swimming pool: the water always looks shallower than it really is.

If the block has a thickness of t, then the optical thickness t_1, is

$$t_1 = t/n \qquad (1.4)$$

So, for the previous example, if the block is 6 mm thick then the optical thickness is only 3.95 mm. This property of refractive index can be put to good use, especially in optical systems with short focal lengths or where the design is

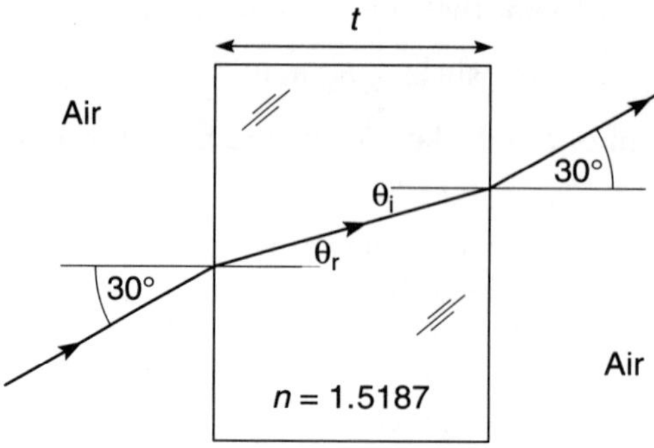

Figure 1.3 Refraction through a plane parallel-sided block of glass

mechanically constrained. For instance, some high-powered microscopes are corrected to operate with a cover-glass between the sample and the lens. This gives the user a chance to view delicate materials without worrying about damaging the sample.

Snell's law applies equally to curved surfaces. Consider the case of a curved surface as shown at Figure 1.4. The surface shown is a section through a spheric; assume that the glass is BK7. The normal at any position on the surface is represented by a radius R, which in this case is 50 mm. For a ray parallel to the optical axis striking the surface at a height, h = 10 mm, the angle of incidence θ_i is given by

$$\theta_i = \sin^{-1}(10/50)$$
$$= 11.5°$$

We now have the following familiar parameters: n_1 = 1.0, n_2 = 1.5187, θ_i = 11.5 °;

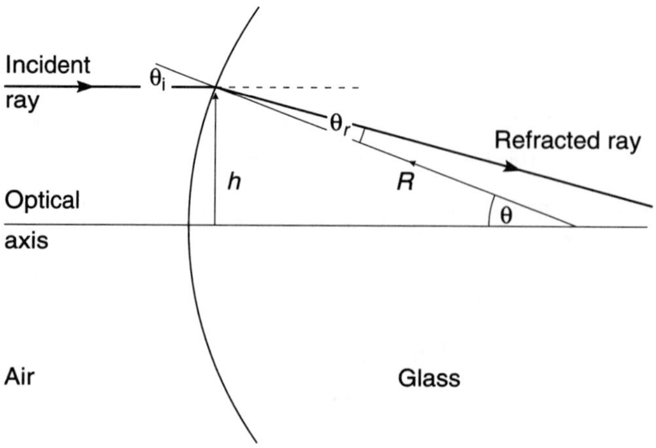

Figure 1.4 Refraction at a curved surface

from them we can calculate the angle of refraction:

$$\begin{aligned}\theta_r &= \sin^{-1}[(n_1 \sin \theta_i)/n_2] \\ &= \sin^{-1}[(1.0 \sin 11.5)/1.5187] \\ &= 7.6\,^\circ\end{aligned}$$

The angle the ray makes with the optical axis is given by $\theta_i - \theta_r$, in this case $11.5 - 7.6 = 3.9\,^\circ$.

This analysis can be continued through as many air-to-glass and glass-to-air interfaces as required, and is the basis for ray-tracing optical systems. Although we have presented this example it is seldom cost-effective to do these laborious calculations in anything other than simple cases; otherwise it is better to use a ray-tracing computer program.

How does a lens work

References to 'light rays' are a useful aid in understanding many optical concepts, but in reality light travels as a wavefront. To understand how a lens works, the concept of a wavefront has to be understood.

Imagine a pebble dropped into a still pond and the ripples running away from the centre of disturbance. Each circular ring emanating from this point source can be thought of as a wavefront, a line of equal phase or equal distance from the centre of disturbance. The rays emanating from the point source will be radial and therefore at right angles to the wavefront. In nearly all real situations light rays are orthogonal to the wavefront, only in some crystals is this not true.

To continue the analogy, over a fixed period the wavefront will expand to a certain radius. If one isolated region of the pond were of treacle rather than of water then, the expanding ripple would slow and the circle would distort. However, the wavefront would still represent a line of equal phase. Thus, the distance travelled by the ray depends on the velocity of the ray in the material and the elapsed time.

Now take this two-dimensional model into three dimensions and consider a region of uniform refractive index. A point disturbance, or light source, will result in the propagation of a spherical wavefront. A wavefront represents a surface with points of equal phase, or time, which leads us to the idea of optical path, given by

$$\text{optical path} = ns$$

where n is the refractive index and s is the physical distance travelled; this simple formula expresses Fermat's principle.

If all rays emanating from a single point at time zero are brought to a focus some time later, then all the rays must have travelled the same optical path length. Clearly, for an optical system constructed from many air–glass interfaces the total optical path of a ray is constructed from the sum of the products of ns or, to put it another way, the integral over the elemental distances δs. Thus

$$\text{optical path} = \int ns \, \mathrm{d}s$$

By way of an example, if a simple biconvex lens forms an image of an object then rays coming from the object to the edge of the lens and then to the image traverse more air and less glass than a ray along the optical axis; however, the refractive index of glass is higher than that of air, so the optical paths equate.

An alternative approach is to consider the spherical wavefront emerging from the point object and striking the lens. The axial part of the wavefront will strike the lens first and be slowed because of the higher refractive index of the glass. At some later time, the extremes of the wavefront will be slowed down by passage through the extreme edge of the lens. Here, then, the axial portion of the wavefront is retarded with respect to that at the extreme, converting the diverging wavefront to one that converges to the image point.

Paraxial optics

Before we move on to discuss the lens equations that enable a basic system to be designed or its performance predicted, we must look at the boundaries within which paraxial optics apply.

Paraxial optics (paraxial means close to the optical axis) is a simplification of what actually occurs in any lens. But for most cases it is a very powerful tool in the outline design of systems. It assumes that rays never deviate far from the optical axis and that aberrations have no effect on the image size. The rules of paraxial optics are:

1. Wavefronts are smooth surfaces of constant phase, or optical path arising from the source.
2. Rays are normal to the wavefront.
3. Energy travels along the rays and the flux at any point is proportional to the ray density at that point.
4. Optical path lengths along all rays between corresponding wavefronts are equal.
5. Diffraction is ignored. (For an explanation of diffraction see page 20 and Appendix A.)
6. $\sin \theta = \theta$, i.e. all angles are small (note that θ is in radians).

Paraxial optics is valid when (a) applied away from the region of focus, (b) not at the edge of a geometric shadow and (c) very much further than 100 wavelengths from the light source.

If light rays deviate through relatively large angles, then errors, or aberrations, are introduced to the wavefront that can destroy imaging performance. The usefulness of paraxial optics decreases as aberrations increase. For example, if paraxial equations are used to determine the focal position of a large biconvex singlet then the result will be useful only for those rays very close to the axis of the lens. In reality, spherical aberration will result in the 'circle of least confusion' or smallest spot size being inside the paraxial focus.

Thus paraxial optics is only an approximation to the real world. Used intelligently, the predictions made within this model will serve the optics user very well. However, lack of understanding of the basic limitations will lead to some surprises.

For a well-corrected optical system – that is, one that introduces only small errors to the wavefront – paraxial optics will yield an accurate prediction of performance. It will also provide a convenient starting place from which to evaluate the system's performance.

Sign conventions

In the analysis or design of an optical system, light is usually considered to travel from left to right, and this convention is maintained in this book. Some authors, for particular reasons, treat light as travelling from right to left, so be sure to check which conventions apply when using formulae from other sources.

We here employ five simple rules:

1. Heights above the optical axis are positive and those below are negative.
2. Distances to the right of a reference point are positive and those to the left are negative.
3. The focal length of a converging lens is positive and a diverging lens is negative.
4. A surface with its centre of curvature to the right has a positive radius and one with the centre of curvature to the left has a negative radius.
5. A negative image distance is a virtual image, i.e. the light will diverge from the lens.

Figure 1.5 will help to cement some of these conventions in your mind. The

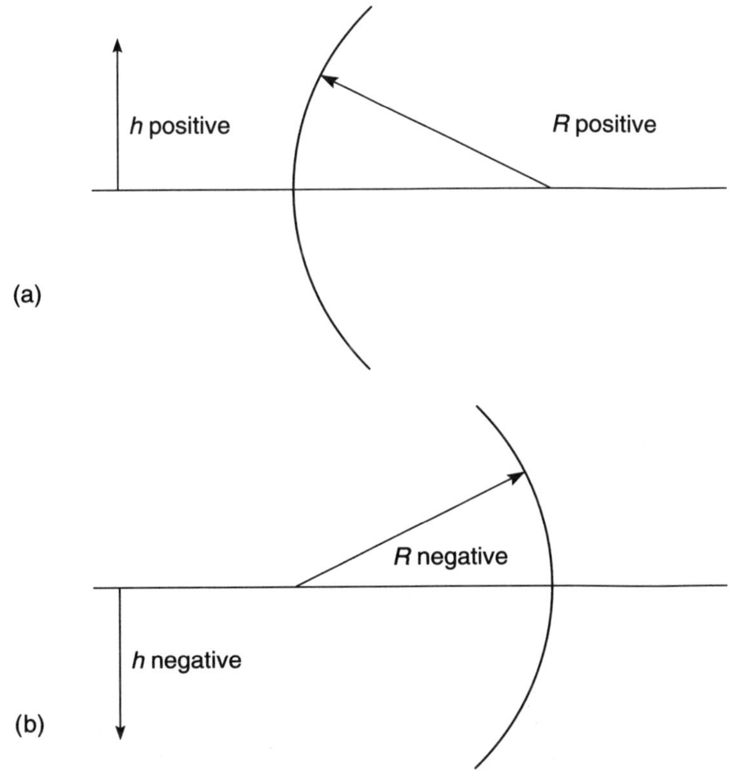

Figure 1.5 Lens and image conventions

conventions will also become clearer and more easily understood as some equations and examples are worked through below.

Lens shapes

A singlet lens can take several different shapes. These are illustrated in Figure 1.6 and are now described:

(a) Biconvex

Positive focal length. Used for symmetrical applications where object and image distances are similar, i.e. low-performance relay lenses, light gatherers for measurement, simple condensers and as basic laboratory tools. Best used where image-to-object ratios are near unity or less than 5 : 1.

(b) Planoconvex

Positive focal length. Used for asymmetric applications where the convex side of the lens should face the long conjugate, thus enabling both lens surfaces to contribute to light bending. Used as low-performance telescope objectives, monochromatic collimators, and in relay systems. Can be used for narrow-bandwidth beam expansion, as for example in laser beam expansion.

(c) Biconcave

Negative focal length and forms only virtual images. Used to diverge collimated light or as part of a beam expansion system, such as the Galilean telescope shown in Figure

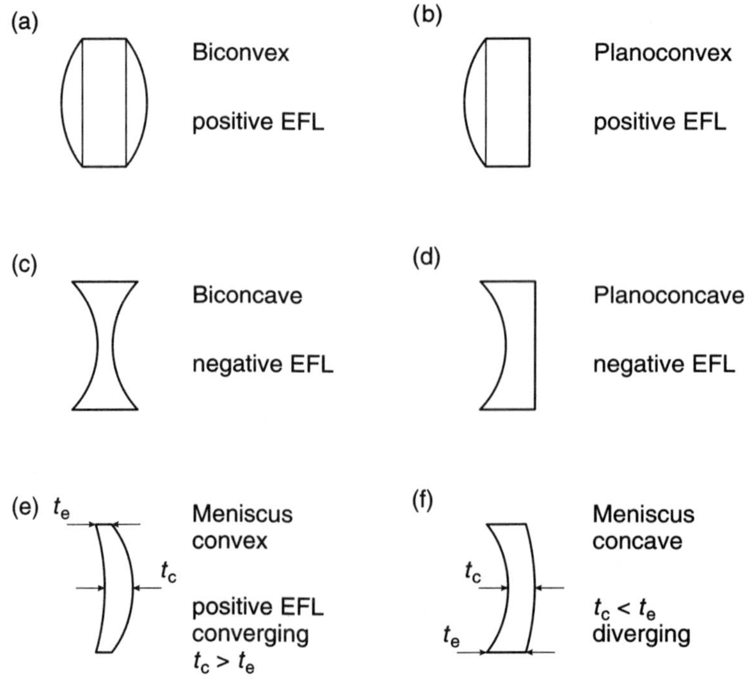

Figure 1.6 Lens shapes

1.9(b). Remember that the most steeply curved face should face the long conjugate distance so that both surfaces contribute to light bending.

(d) Planoconcave

Negative focal length and only forms virtual images. The same comments apply here as for the biconcave lens. There is no sensible rule of thumb that can be applied to distinguish between the various uses of these two types. Often the planoconcave is used in flattening the field curvature, particularly in camera lenses and charge-coupled device (CCD) camera applications.

(e) Meniscus convex (converging)

Positive focal length. Commonly used as corrective elements in systems that employ several elements, for example as a spectacle lens correcting ophthalmic defects or to shorten the focal length of a converging element. In this latter case, the focal length of the combination is shortened while the numerical aperture is maintained. Resolution performance is proportional to f-number (symbolized as *f#*) and hence the meniscus lens can be used to increase resolving power.

The positive meniscus lens is also successfully employed in simple cameras as the only image-forming element; used with the appropriate aperture stop it can give an acceptable imaging performance in this application.

(f) Meniscus concave (diverging)

Negative focal length. As meniscus convex above but not imaging.

The most useful general rule to apply to all lenses is that

> The most steeply curved side of the lens should face the longest conjugate distance

Lens equations

We can now discuss the most useful and often used paraxial equations, together with some limitations.

An important restriction to the following formulae is that they only apply to 'thin lenses'. A 'thin lens' ignores the contributions made to aberrations by the amount of material that the light traverses. For practical purposes a thin lens can be thought of as having a thickness of less than a tenth of its aperture.

For a thin lens the refraction of light is considered to occur at a single plane from which the focal length is measured. The lens maker's formula, or thin-lens equation, can be used to determine focal length:

$$1/f = (n - 1)(1/R_1 - 1/R_2) \tag{1.5}$$

where f = focal length of lens
n = refractive index of lens material
R_1 = spherical radius of first lens surface
R_2 = spherical radius of second lens surface

Consider the example of Figure 1.7 In this case,

$n = 1.5187$, $R_1 = 50\,\text{mm}$, positive according to our sign convention, $R_2 = -50\,\text{mm}$, negative according to our sign convention.

Hence,

$$1/f = (1.5187 - 1)(1/50 + 1/50)$$

(note the change of sign between $1/R_1$ and $1/R_2$),

and $$f = 48.2\,\text{mm}$$

When the thickness of a lens becomes significant, perhaps when its thickness exceeds 20 per cent of its diameter, a more detailed form of Equation (1.5) can be applied. This is the thick-lens formula:

$$1/f = (n - 1)\,[1/R_1 - 1/R_2 + t_c(n - 1)/nR_1R_2] \tag{1.6}$$

where t_c is the centre thickness of the lens.

Referring to Figure 1.7 again, and taking the parameters from the previous example, but adding a thickness t_c, of 5 mm, we can see that the focal length increases to

$$f = 49\,\text{mm}$$

This is an increase of about 1 mm or 2 per cent, a change in focal length that is probably negligible for the first-order layout of a simple system.

Magnification

The magnification of an optical system can be defined in terms of linear dimension change or angular change, and each of these can be expressed in several different ways. Appendix C presents a number of different treatments of the magnification

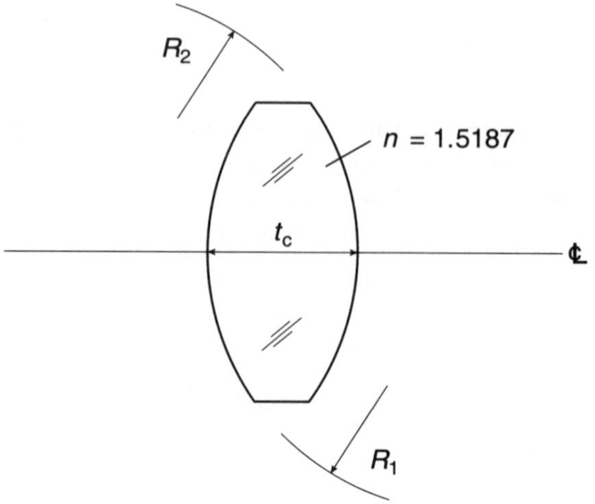

Figure 1.7 Simple singlet

question. Here, for the purposes of illustration, we look at one of the simplest forms of linear magnification expression:

$$m = v/u \tag{1.7}$$

where m = linear magnification
u = distance from object to thin lens
v = distance from thin lens to image

Inspection of Figure 1.8 and the manipulation of the similar triangles will quickly show that:

$$m = v/u = h_2/h_1$$

which is an alternative statement of linear magnification.

The above forms relate to transverse linear magnification, that is, they derive only from dimensions orthogonal to the optical axis. The magnification of an object along the optical axis is different and is expressed by

$$m_L = m^2 \tag{1.8}$$

where m_L is the longitudinal magnification.

We can see the difference between transverse and longitudinal magnification immediately if we consider the different conjugate conditions along the optical axis of an object. This becomes clearer when we consider the next equation.

A very simple relationship between conjugate positions exists for the thin lens, sometimes called the Gaussian lens formula, as follows:

$$1/u + 1/v = 1/f \tag{1.9}$$

Consider as an example our equi-biconvex lens of the previous examples, referring to Figure 1.8 and recalling parameters to which we can add the height h_1 and distance u to form our data set of $f = 49\,\text{mm}$, $h_1 = 10\,\text{mm}$, $u = 130\,\text{mm}$.

To find the second conjugate of the pair we substitute for u and f in Equation (1.9) above, as follows:

$$\begin{aligned} v &= fu/(u - f) \\ &= 78.7\,\text{mm} \end{aligned}$$

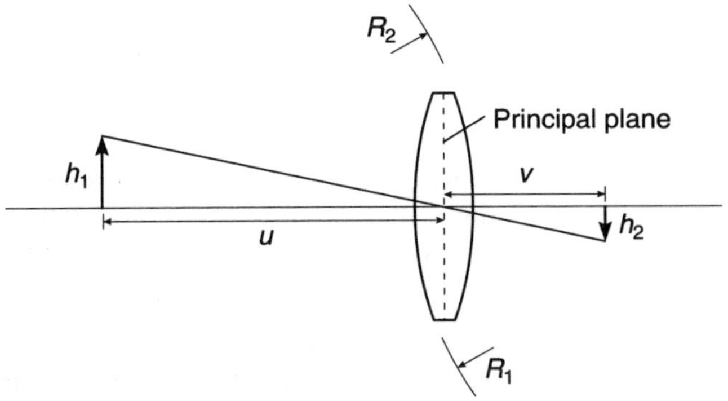

Figure 1.8 Magnification

To determine magnification we can employ Equation (1.7) using our new-found value for v as follows:

$$\begin{aligned} m &= v/u \\ &= 78.7/130 \\ &= 0.61 \end{aligned}$$

The fractional value for magnification represents a demagnification or diminishing of the image with respect to the object.

We can calculate the longitudinal image displacement for a small change in object position by differentiating Equation (1.7) for u and v to yield

$$m^2 \, \mathrm{d}u = \mathrm{d}v \qquad (1.10)$$

All of the above equations refer only to systems where the object and image spaces are of the same refractive index.

Real and virtual images

A real image is one that can be accessed physically. That is, we can place a screen at the focus and an image will be formed. A photographic lens forms a real image on the film of the camera, a relay lens forms a real image and a laser-focusing lens forms a real image.

A virtual image is one that can be observed only by looking through the lens that forms it. A virtual image cannot be accessed by any physical means. Examples include a magnifying glass and the Galilean telescope, which is often used as a laser beam expander. See Figure 1.9 for typical real and virtual images.

Principal planes and nodal points

The thin-lens approach to optical systems design is, of course, only applicable to thin lenses. When lens thickness starts to be appreciable with respect to diameter (>20 per cent), or when a system of lenses is considered, a more sophisticated approach is required.

There is a median between the simplicity of the thin lens and the sophistication of computer-based ray tracing, namely the thick lens. Thick-lens analysis is not limited to a single element, but is equally applicable to a combination of thin lenses or a system of thick lenses.

Thick lenses make use of a model where light bending is considered to occur at two independent planes. These hypothetical planes are usually contained within the body of the lens, but not always. The planes themselves usually exhibit some curvature. See Figure 1.10(a) and (b).

For an optical system in air, the positions where the principal planes cross the optical axis coincide with the nodal points of that system. More exactly, the nodal points of a thick lens are those points of an optical system where a unity-magnification conjugate

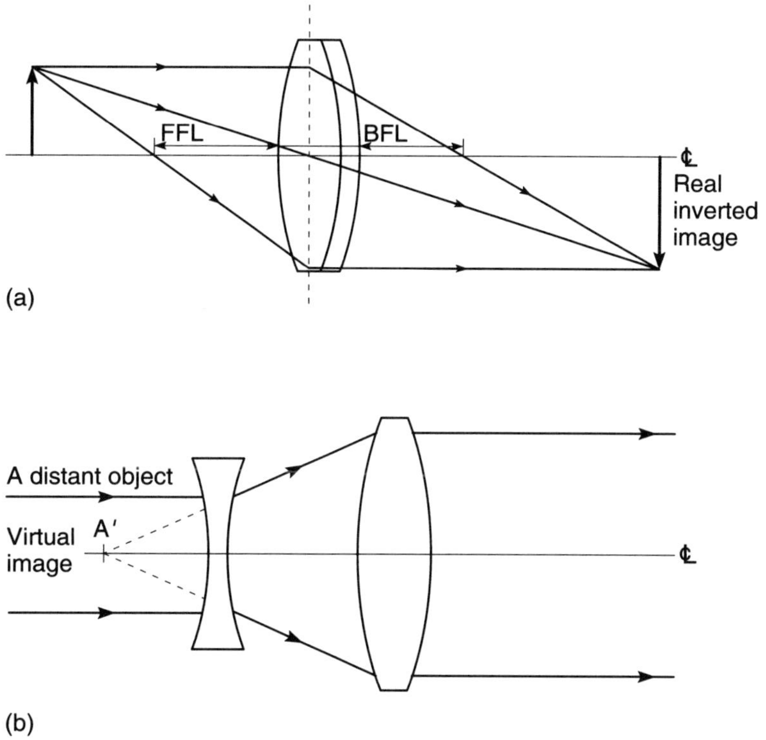

Figure 1.9 Real and virtual images: (a) real image, (b) virtual image

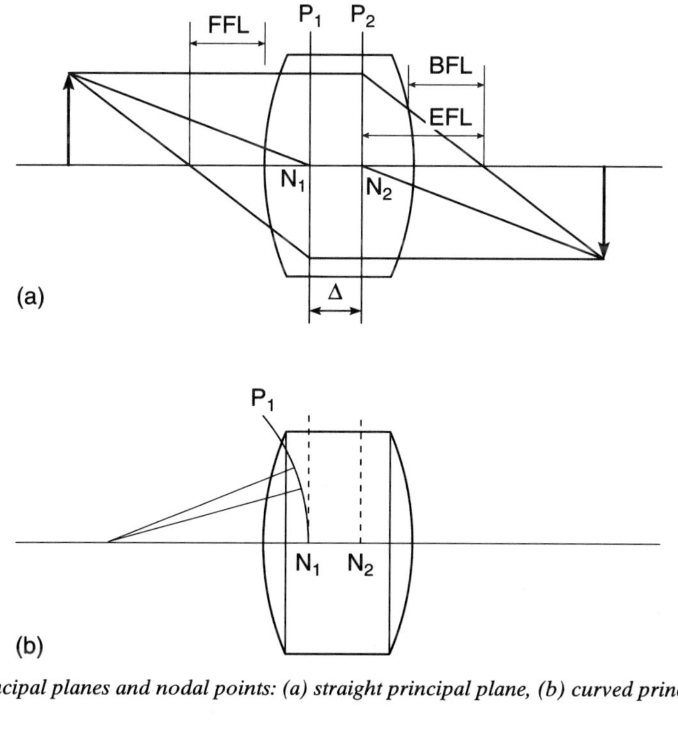

Figure 1.10 Principal planes and nodal points: (a) straight principal plane, (b) curved principal plane

pair exists. The angles of rays passing through the object point and forming the image point are maintained. See Figure 1.11.

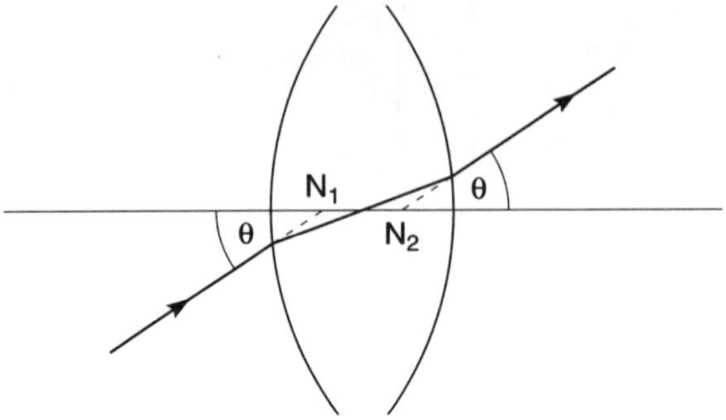

Figure 1.11 Nodal points as a conjugate pair

Thick lenses introduce the idea of effective focal length (EFL) and front/back focal length (FFL/BFL); see Figure 1.10 for the nomenclature that will be applied in this section. The BFL and FFL are measured from the lens vertex to the focus, whereas the effective focal length is measured from the second principal plane to the focus. Nodal points are labelled N_1 and N_2, with corresponding principal planes P_1 and P_2. For a thick lens, conjugate distances (u and v) are also measured from the principal planes.

The principal points, focal positions and nodal points of a lens are known collectively as the cardinal points.

For biconvex or biconcave lenses the principal planes will be located symmetrically within the body of the lens. As a general rule, the principal planes will be separated by one third of the centre thickness of the lens. Figure 1.12 shows typical positions of principal planes for the six types of singlet lens.

Taking together these comments, equations and definitions, we can draw the definitive thick lens as in Figure 1.13.

The distances S_1 and S_2 are given by

$$S_1 = -f(n - 1)t/R_2 n \tag{1.11}$$

$$S_2 = -f(n - 1)t/R_1 n \tag{1.12}$$

Taking our biconvex example of before we can find the distance of the principal planes from the vertices of the lens. With respect to nomenclature and sign convention, if S_1 is positive then P_1 (N_1 in air) is to the right of R_1. Conversely, if S_2 is negative then P_2 (N_2 in air) is to the left of R_2.

The data set here is $n_e = 1.5187$ (lens), $R_1 = 50\,\text{mm}$, $R_2 = -50\,\text{mm}$, $t = 5\,\text{mm}$, $f = 49\,\text{mm}$. Substituting these values in Equations (1.11) and (1.12) gives the following values for S_1 and S_2:

$$S_1 = 1.67\,\text{mm}$$

$$S_2 = -1.67\,\text{mm}$$

In air these positions will represent the principal points and the nodal points of the system.

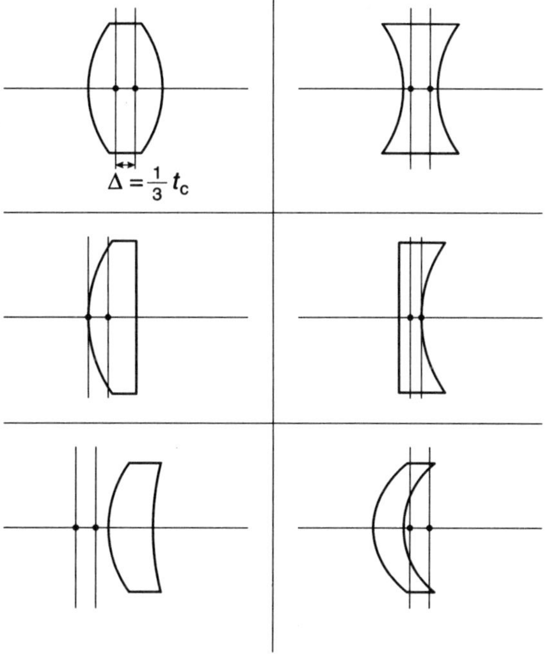

Figure 1.12 Position of principal planes

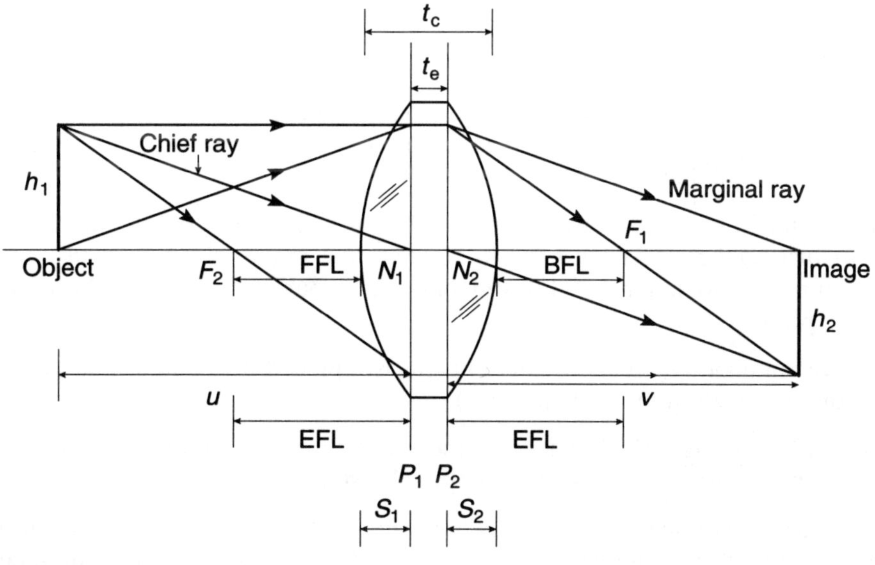

Figure 1.13 Thick lens

Referring to Figure 1.14, rotation of the lens about the nodal point N_2 does not affect the position of the focus. This fact is used by an instrument called a nodal slide, which can be used to locate the position of the nodal points and hence to measure the focal length of complex lenses. Nodal slides are also particularly useful for assessing off-axis performance. Once a lens has been correctly set, i.e. rotation about the second nodal point does not affect image position, then the off-axis image is easily observed through a microscope. If we use star test criteria the lens performance with field can be visually assessed (see 'star test' in Appendix A).

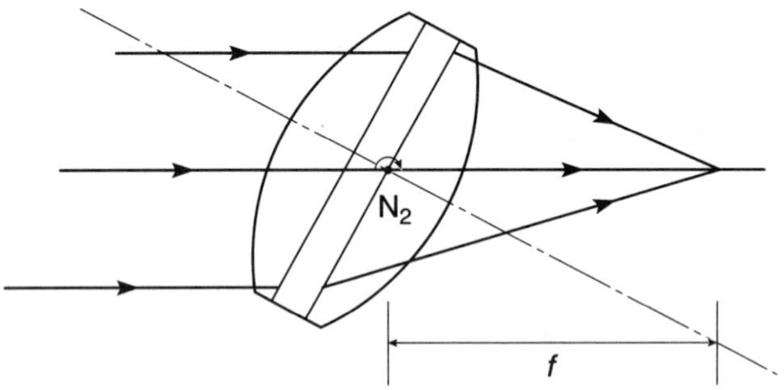

Figure 1.14 *Nodal slide*

Ray nomenclature

Before we move on to the assessment of images and related matters it is useful to review the terms optical engineers employ to describe the basic optical system and the rays used to trace, and hence assess, a design.

Field height, *h*, refers to the height of the object from the optical axis and in most cases is normalized to 1.0. A field height of zero refers to the rays that come from the optical axis to fill the entrance pupil.

The *paraxial* ray, $h = 0$, goes through the centre of the lens along the optical axis. A *zonal* ray starts on the optical axis and enters the lens at 70 per cent of the maximum pupil diameter. A *marginal* ray also starts on axis but passes through the extreme of the entrance pupil diameter.

At a field height of 1.0, the ray that passes through the centre of the stop is called the *chief* or *principal* ray. At least two other extreme marginal rays are usually traced originating at field height of 1.0 and passing through the top and bottom of the entrance pupil.

In infinite conjugate systems the field height is replaced by angular field of view (not to be confused with numerical aperture).

Figure 1.13, showing a thick lens, uses these basic rays, which may be considered the minimum necessary to characterize an optical system.

As long as the optical system is symmetric, centred and without tilts it is only necessary to trace rays in one axis. But this is not true for off-axis imaging such as the example shown in Figure 1.15. In this case, rays that lie in the plane containing the object and optical axis are defined to be *tangential.* The rays in a plane at 90° to the tangential are called *sagittal.* These definitions ensure greater asymmetry in the tangential ray angles than in the sagittal, a fact that becomes especially important in dealing with coma and astigmatism. Any ray that arises outside the tangential plane is referred to as a *skew ray.*

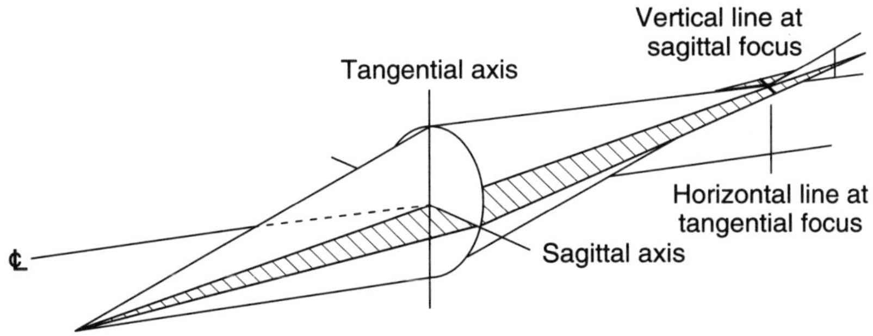

Figure 1.15 Tangential and sagittal axes

The fact that the tangential axis is always defined in terms of the object plane means that tangential aberration, in symmetric designs, is always greater than sagittal aberration, usually by about 3 to 1.

Lens nomenclature

Lenses too have their own nomenclature, terms and definitions, which are used in the assessment and design of optical systems.

The *stop* of a lens is its smallest aperture and sets a limit to the height of the marginal ray. If you look into any complex, non-symmetric lens from each end in turn, the stop diameter will be greater or smaller depending on the viewing direction. This is because what you see are the images of the stop through different lens elements. These images are called the *entrance pupil* and the *exit pupil.*

Stop and exit pupil position are illustrated in Figure 1.16, along with the marginal rays and chief ray of the system. The position of the pupil is located where the chief ray crosses the optical axis, and the diameter of the pupil is set by the height of the

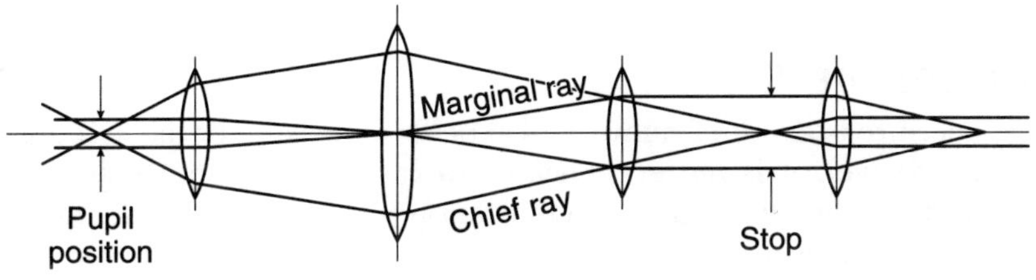

Figure 1.16 Exit pupil and stop

marginal rays at this point. The example shown is an erecting eyepiece, where obviously the pupil position is crucial to the comfort of the user. But often the existence of the pupil is ignored by the unwary, and the clear diameter of the lens is taken to be the limiting factor in accepting the light. This can be very wasteful, particularly in projection designs, where carefully matching the numerical aperture of the condenser to exactly fill the entrance pupil is vital in maintaining luminous efficiency.

In a thin lens the difference between the respective positions of the stop and of the entrance pupil can be ignored and the aperture is effectively set by the clear glass diameter.

f-number

Most photographers are familiar with the f-number settings on their cameras and the relationship between lighting and the chosen setting. As the scene becomes darker the 'speed' is increased by enlarging the stop diameter, thus decreasing the f-number setting.

The f-number (symbolized as $f\#$), for systems with one infinite conjugate, is calculated from the simple ratio

$$f\# = f/\phi \tag{1.13}$$

where f is the effective focal length (EFL) of the lens and ϕ is the entrance pupil diameter of the lens, or, in the case of a thin lens, its clear diameter.

This relationship between diameter and focal length sets the largest angle for the marginal ray through the lens, σ_r, so that

$$\tan(\sigma_r) = \phi/(2f) \quad \text{or} \quad \frac{1}{2f\#} \tag{1.14}$$

The f-number is therefore a measure of the energy collection capacity of a lens, being proportional to the area of its entrance pupil and therefore to the square of its radius. For example, a lens with a focal length of 35 mm and a clear diameter of 25 mm is a f1.4 lens and is relatively 'fast'. It will have a marginal ray angle of 19.65° and produce an image four times as bright as that from a f2.8 lens.

In a simple lens, the clear diameter is equivalent to the entrance pupil, but this is not true for more complex lenses or aspheric singlets. For these lenses, the limiting rays through the optical train are set by the entrance pupil diameter and the aperture stop. It is a common mistake to assume that all of the clear diameter of the lens is used in calculating the f-number; this is not the case.

Numerical aperture (NA)

Numerical aperture is akin to f-number, also being a measure of the acceptance angle of the lens. However, it is more applicable to cases where both object and image distances are finite. In such a case, magnification m, must be considered. Numerical aperture is calculated using

$$\text{NA} = m\phi/[2f(m+1)] \tag{1.15}$$

At high magnifications, $m/(m+1)$ approaches unity, and the equation simplifies back to

$$\mathrm{NA} = \phi/2f \text{ or } \frac{1}{2f\#} \tag{1.16}$$

Numerical aperture is also used to define the acceptance or emission angle of an optical fibre. In this case the NA is calculated from

$$\mathrm{NA} = n_0 \sin(\sigma_r) \tag{1.17}$$

where σ_r is the half-cone angle of the emitted light,

Or

$$\mathrm{NA} = (n_1^2 - n_2^2)^{1/2} \tag{1.18}$$

where n_0 = refractive index of medium into which light is being emitted
n_1 = refractive index of core
n_2 = refractive index of cladding.

Power

The power of a lens is the inverse of its focal length in metres and is most commonly used in the ophthalmic field for specifying spectacles. The units of power are dioptres.

$$\text{power} = 1/\text{focal length} \tag{1.19}$$

thus a 5-dioptre lens must have a focal length of 200 mm.

Spherometers, instruments for measuring the radius of a spherical lens surface, often provide the measurement in dioptres. In this case the radius, R, measured in millimetres, is not the inverse of the dioptre reading, D, but must be calculated from the formula

$$R = 525/D \tag{1.20}$$

Here 525 is derived from the formula $1000(n-1)$, which assumes glass to have an index n of 1.525.

Resolution

Resolution is a measure of the smallest feature an optical system can reproduce or discriminate. It is usually expressed as a modulation depth at a spatial frequency, typically in line pairs per millimetre (lp/mm) or sometimes lines per millimetre (l/mm), which is more common in the world of TV broadcasting. For more demanding applications, such as television camera lenses, the resolution should be specified at two or three field heights, and, in the sagittal and tangential planes, spectral bandwidth may also be required. In certain instances, usually involving telescopes, eyepieces or spectacles, angular resolution is quoted in line pairs per milliradian (lp/mrad).

The most important factor affecting resolution is the size of the image, which is ultimately limited by diffraction and more often limited by aberrations. A discussion of these matters follows.

Diffraction

It is common practice to treat the movement of light through any medium as a collection of rays which move only in straight lines. For many applications this is a perfectly acceptable approximation. In practice, though, some light will always be spread out beyond the area expected from linear ray tracing; this is due to diffraction.

Diffraction occurs because of the wavelike behaviour of light, which causes each point on a wavefront to behave as a secondary source. The secondary wavelets interfere, producing an interference pattern whose intensity quickly falls off outside the angle of the initial direction of propagation.

The two regimes of diffraction are:

- *Fresnel:* concerned with diffraction close to the diffraction source.
- *Fraunhofer:* concerned with diffraction at an infinite distance from the diffraction source. A lens that forms an image of an infinite source at its focal plane is giving the far field diffraction pattern of the aperture, and so is an example of Fraunhofer diffraction.

Most lenses are circular, and, provided that the aperture is evenly illuminated, the intensity of the diffraction pattern can be described by

$$I_{(z)} = I_0[2J_1(z)/z]^2 \tag{1.21}$$

Where I_0 is the peak intensity at the centre of the image and $J_1(z)$ is a Bessel function of the first kind, zero-order, defined by the expansion

$$J_1(z) = \sum_{n=1}^{\infty} (-1)^{n+1}(z^{2n-2})/(n-1)!n!2^{2n-1} \tag{1.22}$$

and $z = \pi D \sin\theta/\lambda$. In this case D is the aperture diameter and θ the angular radius from the centre of the interference pattern.

For a slit aperture the intensity pattern is given by

$$I_z = I_0[\sin c\ (z)]^2 \tag{1.23}$$

where $z = \pi w \sin\theta/\lambda$ and w is the slit width.

For both a slit and a circular aperture, the assumption of even illumination does not apply to a laser beam. The Gaussian distribution of a helium–neon laser beam, for example, modifies the diffraction pattern. Under normal circumstances, a Gaussian distribution is transformed into another Gaussian whose profile must be determined using the formula for beam waist and wavefront radius of curvature in Chapter 8.

Airy disc

In simple geometric optics, collimated light passing through a perfect lens would be brought to a point focus, the image would be infinitely small and there would be no limit to resolution. In reality, the lens will produce an interference pattern (see Chapter 8), comprising a bright central region surrounded by alternate dark and bright rings. This is the Airy disc and is caused by diffraction. The intensity distribution of this pattern is described by a first-order Bessel function in which 84 per cent of the light is in the central region and about 7 per cent in the next bright ring.

A lens capable of producing an Airy disc is substantially free from aberrations and is referred to as diffraction-limited. The smallest image that can be resolved will be limited by the radius of the central part of the Airy disc, the size of which in micrometres can be calculated from

$$A_r = 1.22\,\lambda f\# \tag{1.24}$$

For example: an f1.4 lens operating in green light, at 540 nm, would have a diffraction-limited radius of 0.922 μm. Needless to say, not all systems require this level of performance and even fewer can afford to pay for it.

Note that the diffraction limit formula for a coherent Gaussian laser beam is slightly different; see Chapter 8 under 'Lasers'.

Modulation transfer function

Modulation transfer function (MTF) is a vital and common tool in describing the resolution capability of an optical system. It is really a measure of the amplitude response that a lens can produce over a range of spatial or angular frequencies.

Suppose an aberration-free lens is used to image a pattern of alternate bright and dark lines. The image of each point on a bright line will have its own Airy pattern, which will cause the bright lines to spread out into the dark parts of the image. If a detector, smaller than the line images, is scanned across the image plane then the maximum and minimum values can be used to calculate the contrast or modulation depth from

$$\text{Contrast} = (I_{max}-I_{min})/(I_{max}+I_{min}) \tag{1.25}$$

By altering the spatial frequency of the pattern, a plot of modulation depth against frequency in the image plane can be drawn.

Strictly speaking, what has been described is a square-wave response or contrast transfer function (CTF), while MTF is the sine wave amplitude response of a lens. It is often easier to measure the CTF, and so it is useful to be able to convert between the two functions using Equations (1.26) and (1.27):

$$\text{MTF}(v) = (\pi/4)[C(v)+C(3v)/3-C(5v)/5+C(7v)/7\ ...] \tag{1.26}$$

$$\text{CTF}(v) = (4/\pi)[M(v)-M(3v)/3+M(5v)/5-M(7v)/7+M(9v)/9] \tag{1.27}$$

where $C(v)$ is the CTF at frequency v and $M(v)$ is the MTF at this frequency. Note that the $C(9v)/9$ term has been omitted in Equation (1.26) because it is zero.

One of the major advantages in using the sinewave-based MTF is that the system MTF is the product of the MTF of each component. But be careful – this applies only when a lens forms an image on a detector such as a CCD camera. In other circumstances, when for example a relay lens is used with an enlarger lens to produce an image with two object planes, the MTF can only be cascaded if the lenses are separated by a diffusing material. This is because the aberrations in each lens are independent of each other, so they might cancel, improving the MTF, or they might add.

The biggest problem with MTF is often in deciding what is tolerable or sensible to specify for a system. Table 1.2 gives some general indications of the depth of modulation at the specified frequency for some typical applications.

Table 1.2 Depth of modulation for typical applications

Photocopier lens	80% at 5 lp/mm
CCTV lens (f3.5)	On axis 50% at 50 lp/mm (20% at 100 lp/mm)
Camera lens	50% at 25 lp/mm
High-quality CCD lens	50% at 70 lp/mm
Microfilm lenses	80% at 30 lp/mm

Diffraction-limited MTF

Principles of Optics by Born and Wolf (see Further reading) has an excellent description of how to calculate the MTF of a real lens. Nowadays this labour-intensive procedure is unnecessary because of the large number of lens design software packages available that can do the task automatically. However, it is useful to know what the limitation on depth of modulation is likely to be, and this can be done by calculating the diffraction-limited MTF for an aberration-free circular lens in incoherent light using

$$\text{MTF} = (2/\pi)[\cos^{-1}(y) - y(1-y^2)^{1/2}] \tag{1.28}$$

where $\cos^{-1}(y)$ is in radians not degrees

f = spatial frequency

$y = \lambda(f\#)f$

(cutoff frequency) $f_0 = 1/\lambda(f\#)$

An illustration of this response is shown in Figure 1.17, along with the equivalent amplitude transfer function for a coherent light source (see Chapter 8). The two are markedly different, with the partially coherent condition lying somewhere between these extremes. The greater the coherence of a source, for example with a laser or a very small pinhole, the better the modulation depth at lower frequencies, but the high frequency response is reduced.

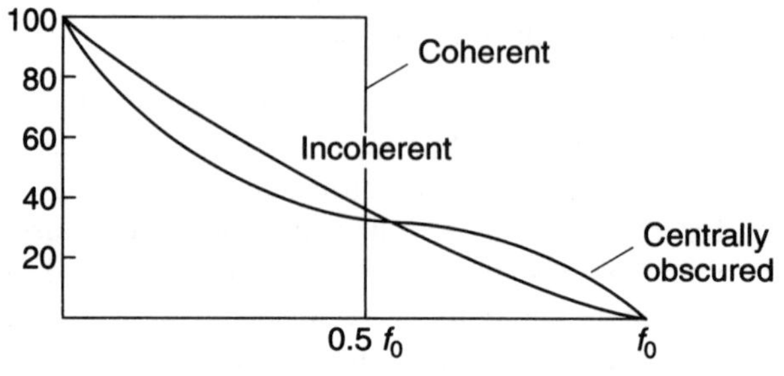

Figure 1.17 MTF curves

The converse of this situation is to have a central obscuration in the lens, in which case the MTF rises above the diffraction limit at high frequencies, but is reduced at low frequencies. Such systems are common in astronomical telescopes and some compact photographic lenses.

Line spread function

At its simplest, the line spread function is a measure of the power distribution across the centre of a line of light. Its importance lies in the fact that its Fourier transform yields the MTF.

Modern optical design software computes MTF by tracing numerous rays through the lens to create a spot diagram. The line spread function is then derived and the Fourier transform taken. Most software packages allow the option of calculating either the diffraction MTF (DMTF), which includes the effects of diffraction, or the geometric MTF (GMTF), which neglects diffraction and is therefore an approximation. The choice of which is most appropriate must be made with knowledge of the likely performance of the lens. A lens that is substantially aberration-free should be assessed using diffraction MTF. Alternatively, if aberrations beyond λ exist then the GMTF can be used to give results that are often a closer match to what will actually be achieved in production.

Depth of focus

The idea of depth of focus (DOF) comes from the assertion that the image plane of any optical system can be moved a small distance in either direction without unduly degrading the image that is formed. Moving away from the plane of best focus will obviously degrade the image and result in a blurred spot, but if the blur cannot be resolved by the detector in use then such image degradation may be deemed indistinguishable from the 'perfect' focused image.

To some extent the size of the DOF depends on the criteria used to define 'acceptable' image degradation. As with resolution limits, DOF depends upon the detection scheme in use and the quality of the lens. There are two methods that the engineer may apply in determining the depth of focus.

Aberration free depth of focus

A simple example of deriving the depth of focus is shown in Figure 1.18. Here the blur circle (blur spot), diameter B, is related to the depth of focus, d_1 and d_2 on either side of the image, so that

$$d_1 = vB/(D-B) \qquad \text{and} \qquad d_2 = vB/(D+B) \tag{1.29}$$

Obviously if the lens is stopped down so that D equals B then the far DOF (d_1) goes to infinity and the near DOF becomes $v/2$. The problem still remains in deciding what the acceptable blur spot diameter is going to be, and this depends critically upon the detection scheme at the image plane. By way of example, suppose that we are using a CCD camera with a 35-mm diameter, f3.5 lens. If we decide that anything below a pixel height, 10 μm, will not be resolved, then d_1 and d_2 are both about 0.035 mm.

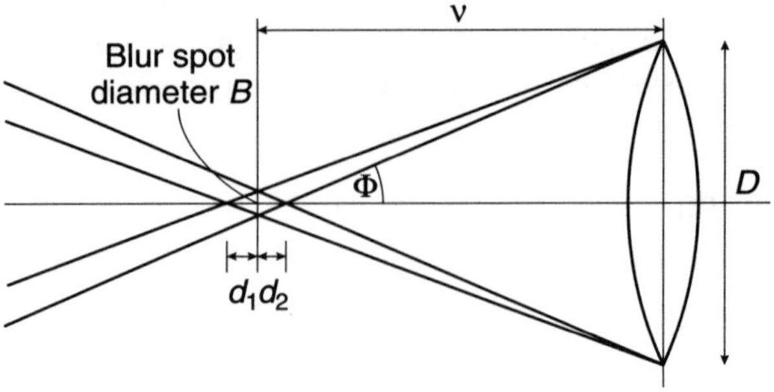

Figure 1.18 Depth of focus

Geometric depth of focus

There are several simplifications in the derivation of the depth of focus above, but generally the results are representative provided the lens is substantially aberration-free. However, when the frequency response complicates the choice of blur spot diameter, then geometric DOF proves to be a very powerful alternative. It is found by transforming the focused cone of light into frequency space, where it is described by a Bessel function of zero order, first kind (see Equation 1.22).

In Figure 1.19, the depth of modulation has been plotted against the normalized

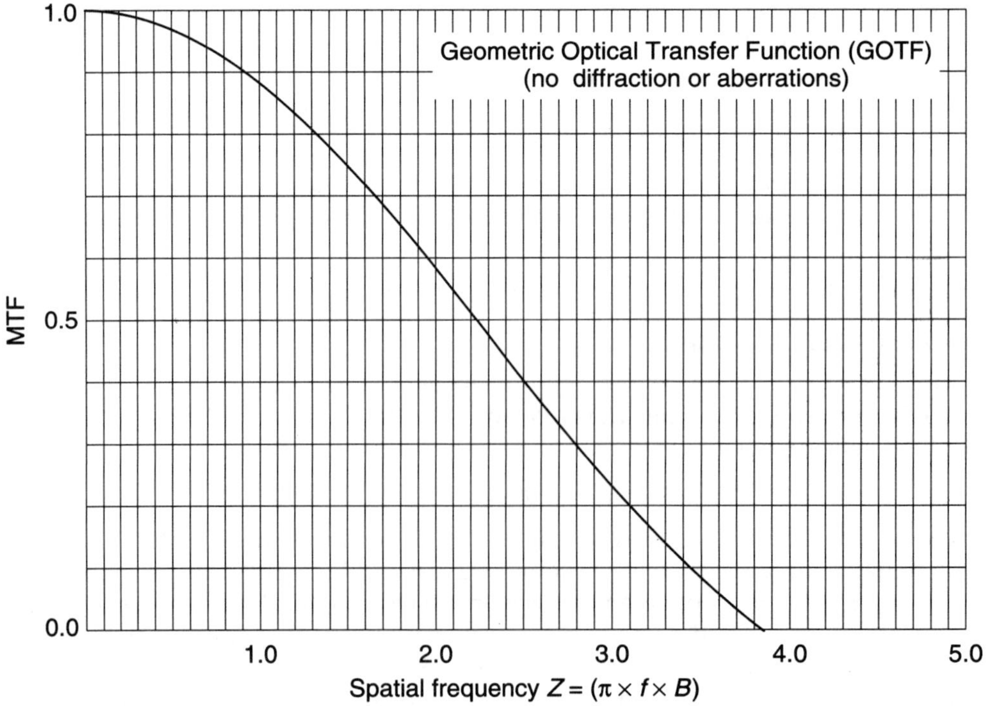

Figure 1.19 Geometric depth of focus

spatial frequency, Z, as given by

$$Z = \pi f B \tag{1.30}$$

where f is the spatial frequency (lp/mm) and B is the blur spot diameter. Now let us suppose that the optical system requires a 50 per cent modulation at 38 lp/mm. From Figure 1.19, 50 per cent modulation occurs at z=2.2. Therefore

$$B = 2.2/\pi \times 38 = 0.018\,\text{mm}$$

and

$$\text{DOF} = \pm\,(B \times \text{f-number})$$

so for an f5 lens the DOF is ± 0.09 mm.

Aberrations

Aberration is the collective term given to the bain of all optical designers, and encompasses all the factors that conspire to prevent us from producing a perfect image from every lens!

Aberrations can be described theoretically using a polynomial expression. A simple lens may be adequately characterized by the use of only the first terms in the expression (first-order). More complex aberration characteristics may require third-order, fifth-order or even higher-order terms. However, for many applications a knowledge of first-order performance is sufficient. The following section is confined to first-order systems and describes the dependance of the individual aberrations on ray height (y) or object height (h).

In reality the limitations of detection and economy prevent it being necessary to eliminate every aberration. So it is important for the engineer to understand what effect each aberration will have, and how much can be tolerated during the design of an optical system.

Spherical aberration

Dependent on y^2

As the ray height at the pupil increases, the rays are focused closer to the lens, which spreads the image at the paraxial focal plane. Rays close to the optical axis land on or near the paraxial focus, but those landing further up the pupil are increasingly defocused. This leads to the concept of a circle of least confusion which 'appears' to be the optimum focal plane for on-axis light, even though it bears no relationship to the paraxial focus. This position is of great importance in radiometric or photometric applications in which light gathering is more important than image quality. A typical example of spherical aberration together with the circle of least confusion is shown in Figure 1.20. Note that most optical design software will calculate the position for optimum resolution as a defocus from the paraxial focus.

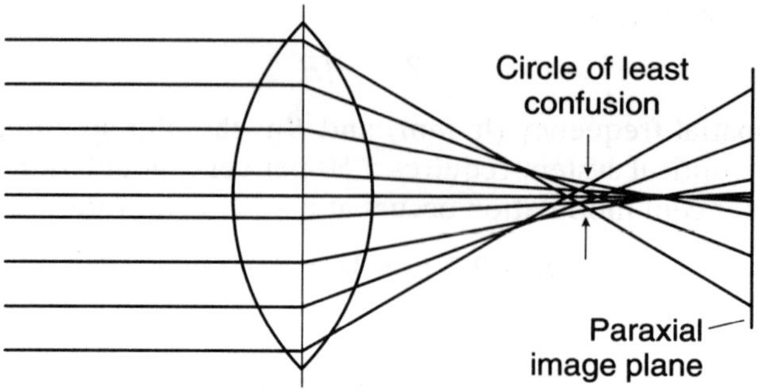

Figure 1.20 Spherical aberration

Spherical aberration can be described as a longitudinal measurement (LSA) or, more commonly, a transverse one (TSA), where

$$\text{TSA} = \text{LSA} \times \tan u$$

u being the marginal ray angle.

Aberrations are often presented graphically in the forms shown in Figure 1.21. The x-axis refers to the relative pupil height about the optical axis, while the y-axis denotes the size of the aberration.

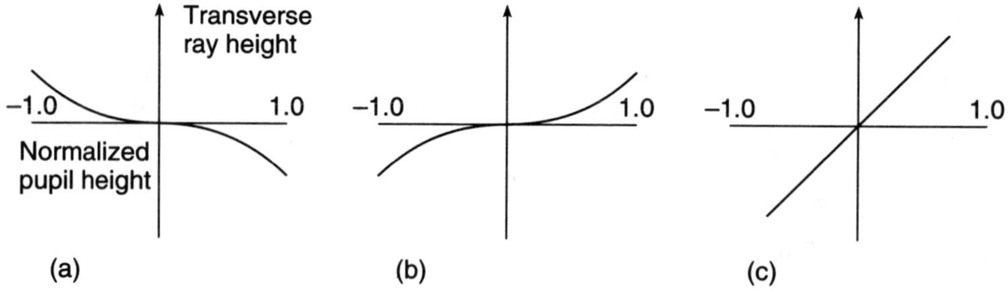

Figure 1.21 Aberration plots: (a) undercorrected, (b) overcorrected, (c) defocus

First-order spherical aberration has a square-law dependence on aperture size, so it can be rapidly reduced by stopping down the lens aperture. Third- and fifth-order types are less readily dealt with, examples of these different orders of spherical aberration being shown in Figure 1.22. Spherical aberration is also dependent on shape and magnification; in general:

Positive lenses have undercorrected spherical aberration

Negative lenses have overcorrected spherical aberration

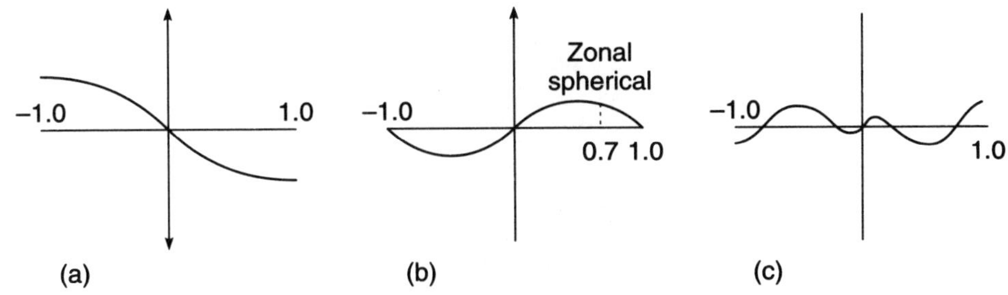

Figure 1.22 Spherical aberration: (a) first-order, (b) third-order, (c) fifth-order

A positive lens operating at unity magnification will have a minimum of spherical aberration; alternatively, computer-optimized doublets are now available with diffraction-limited on-axis performance at single wavelengths. Both types of lens are particularly useful for laser applications.

A further alternative to combining lenses is to use an aspheric lens. This comprises at least one surface, which is defined by the departure from a base radius by a polynomial function. The base radius fixes the paraxial focal length, and the higher-order terms are chosen to minimize spherical aberration.

The aspheric equation is often expressed in terms of the sag height of a surface:

$$X = \frac{cv \cdot p^2}{1+[1-(1+k)cv^2p^2]^{1/2}} + a_4p^4 + a_6p^6 + a_8p^8 \tag{1.31}$$

where cv = curvature
k = conic constant = − (eccentricity)^2
p = $(y^2+z^2)^{1/2}$

Although aspheric lenses can be bought from many sources, it is seldom possible to obtain enough information to ray-trace them during the design phase. Even after practical testing of the lens it can be difficult to decide on performance improvements. A useful starting-point in this case can then be to use the following formula. Set $k = -1$; then

$$A_4 = \tfrac{1}{8}\,cv^3 \qquad a_6 = \tfrac{1}{16}\,cv^5 \qquad a_8 = \tfrac{5}{128}\,cv^7 \qquad a_{10} = \tfrac{7}{256}\,cv^9$$

This will create a parabolic, aspheric surface, which is often a closer approximation to the real performance than is a simple spherical surface of the same radius. The suggested formula is also a useful starting-point in the design of aspherics prior to optimization.

Coma

Dependent on h and y^2

Coma results from the failure of off-axis rays to all arrive at the same image point. Typically, this results in the image becoming smeared and comet-like in shape, as for example in the spot diagram shown in Figure 1.23.

Figure 1.23 Coma spot diagram

Coma is often the least-acceptable aberration, because of the asymmetric distribution of the light. Even if the spherical aberration is reduced to zero, as for an aspheric lens, coma can still be present when an off-axis object is imaged.

A typical aberration plot with coma is shown in Figure 1.24. For a lens with third-order aberrations the sagittal coma is one-third the tangential coma. Once coma is present in a lens it can only be corrected by adjusting the stop position and diameter to vignette extreme rays, thus causing a loss in light levels.

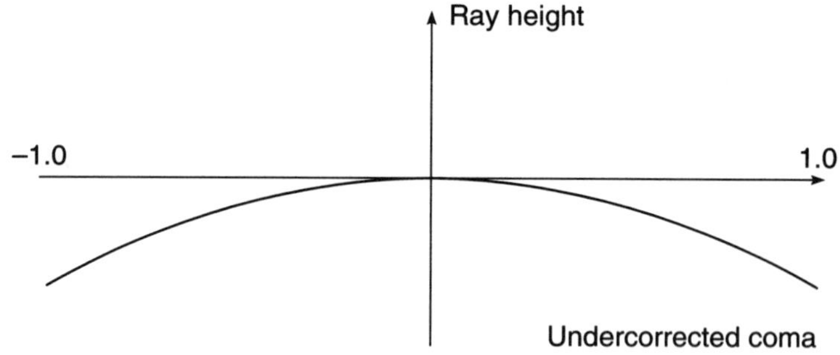

Figure 1.24 Coma

Astigmatism

Dependent on h^2

For an off-axis object imaged by a positive lens, the departure from sine convention (i.e. sin θ does not equal θ) gives rise to a different focal position in the tangential and sagittal planes. Referring to Figure 1.15, it is clear that the tangential image is closer to the lens than the sagittal image. At the tangential image plane the sagittal image is defocused, giving a horizontal line, whereas at the sagittal image plane the tangential image is a vertical line. In between these positions the image is either a circular or an elliptical blur. Astigmatism will influence the choice of focal plane for a detector like a CCD camera or a photographic plate, and it may be necessary to compromise axial performance to reduce the effects of astigmatism.

Astigmatism is easily detected in MTF plots by the separation of the tangential and sagittal curves. As with coma, third-order theory means that longitudinal tangential astigmatism should be three times greater than the sagittal value. When viewing real images, astigmatism can be detected by looking at the image quality of horizontal and vertical bars at the same nominal object height.

Astigmatic sources do exist, and probably the most common is the laser diode, where inherent astigmatism can vary from extremes of 40 μm to about 5 μm. In the plane perpendicular to the output stripe, the light appears to originate from the output facet, while, in the plane parallel to the output stripe, light appears to originate from inside the chip. This can lead to problems when very small images are required, or when a high degree of wavefront correction is needed. Astigmatic sources can be corrected by the use of either a positive or a negative cylindrical lens which adds or subtracts power from one axis. The net effect is to make the apparent object positions in the two axes coincide.

Distortion

Dependent on h^2

If an off-axis image is formed at a different height from that expected from the system magnification then the image is distorted. Positive, or pincushion distortion, transforms a square object into an image with elongated corners. Negative, or barrel, distortion, produces an image with compressed corners; for examples of both types see Figure 1.25.

It is worth remembering that distortion is reversed on either side of the lens, so that pincushion in object space is barrel in image. This reversibility is used to good effect

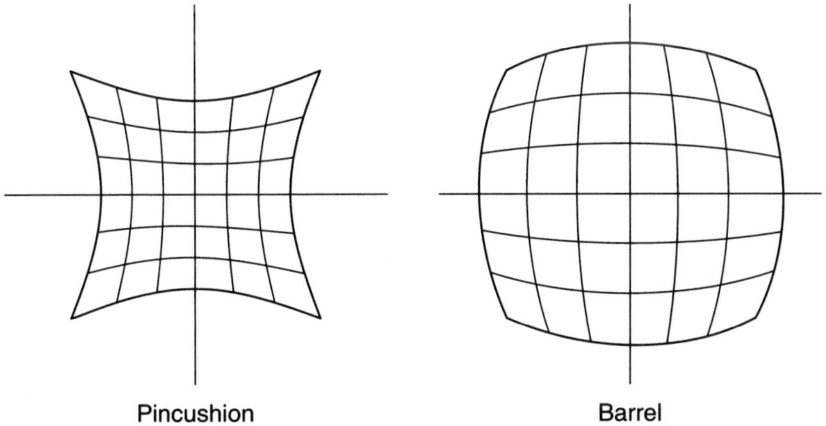

Figure 1.25 Distortion

in CRT projection, where the distortion of the lens can be entirely corrected by introducing distortion onto the CRT. Alternatively, any distortion imposed on a photographic slide can be eliminated by using the same lens in the projector.

Normally, distortion is defined as a percentage of the expected paraxial image height, (h'):

$$\text{distortion} = h'_{\text{actual}}/h'_{\text{paraxial}} \tag{1.32}$$

The amount of distortion that can be tolerated greatly depends on the application; a few examples are given in Table 1.3.

Table 1.3 Examples of acceptable distortion

Photocopying	1–2%
Film projection	2–3%
Metrology	<1%

Distortion-free operation can be achieved by using a well-corrected lens at unity magnification; in this case, coma will also be zero.

TV and film distortion

In the TV and film industry it is commonplace to define distortion in terms of the percentage deviation that can be tolerated in an image. In this case the paraxial image height has no relevance; only the actual picture size and aspect ratio are important. Distortion is now calculated as the percentage difference between the maximum and minimum vertical and horizontal dimensions.

For example, a 4:3 TV picture is projected on a screen to give the picture shown in Figure 1.26. The image has a 2 per cent distortion at a diagonal height of h = 1.0.

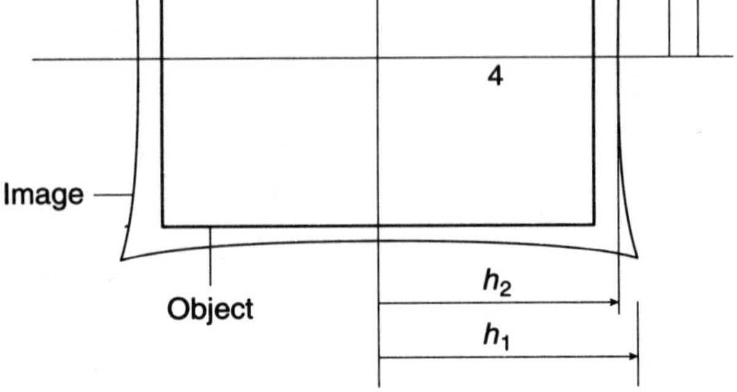

Figure 1.26 TV/film distortion

Normal distortion for a 4:3 image would be as follows:

- Horizontal width, $h = 0.8$, will have a distortion of $0.8 \times 0.8^2 \times 0.02/0.8 = 1.28$ per cent.
- Vertical height, $h = 0.6$, will have a distortion of $0.6 \times 0.6^2 \times 0.02/0.6 = 0.72$ per cent.

In TV/film distortion terms,

- Horizontal distortion = $(h_1 - h_2)/h_2$ or $(4.08-4.0512)/4.0512 = 0.71$ per cent.
- Vertical distortion = $(v_1 - v_2)/v_2$ or $(3.06-3.0216)/3.0216 = 1.27$ per cent.

For such an image the vertical distortion defined in TV terms is more stringent, but equally it is more applicable to the qualitative assessment of the picture made by a viewer.

Field curvature

Dependent on h^2

If we take a simple positive lens and trace rays from a variety of off-axis object positions, we see that the apparent image position is not a plane but in fact a spherical surface curving towards the lens. This result is hardly surprising, because the power of the lens does not change as we move off-axis and therefore the image position is bound to get closer to the lens as the chief ray angle increases. This is demonstrated in Figure 1.27.

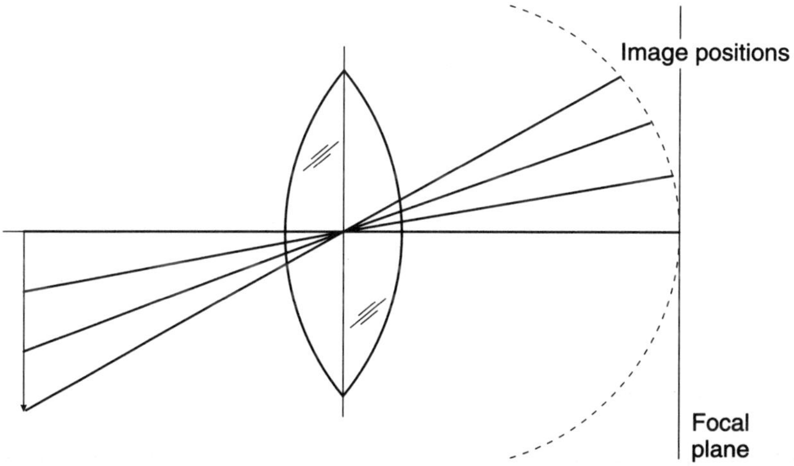

Figure 1.27 Field curvature

Usually, positive lenses have inward-curving or undercorrected field curvature. Negative lenses usually show backward-curving or overcorrected field curvature.

The term 'Petzval curvature' is used to describe the amount of field curvature to be expected from any optical system. It is the curved surface on which images without astigmatism would lie. For a simple thin lens, the longitudinal difference between the

Petzval surface and a plane image surface is given by

$$(hm)^2/2f\# \qquad (1.33)$$

Since it is usually easier to correct lens performance on-axis, and more difficult at the edge of the field, even without field curvature, it is common practice to compromise on-axis performance by defocusing the image plane. For example, the plane of best focus might be taken to be the point at which the zonal field is in focus, i.e. 50 per cent of the lens area is imaging behind focus and 50 per cent in front. Usually lens design software will assist in choosing the plane of best focus for only one field position. Under such circumstances it is a matter of judgement to select the best compromise position, and much depends on the system application.

Longitudinal chromatic

Dependent on the dispersion of the glasses used

We have already seen how the refractive index of a material changes depending on the wavelength of the light (see Equation (1.2)). Usually the shorter the wavelength the higher the refractive index, although glasses with anomalous dispersion do exist. For a simple positive lens this normally means that the blue focal length is shorter than the red. This is illustrated in Figure 1.28. Longitudinal chromatic aberration is the distance along the optical axis between the two extreme foci, and only refers to images on the optical axis.

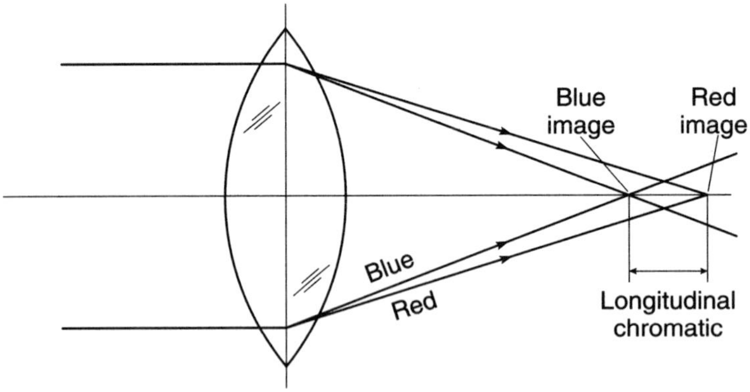

Figure 1.28 Longitudinal chromatic

Correction of chromatic aberration is usually accomplished with an achromatic doublet which combines the different dispersion of crown and flint glasses with different lens powers and shapes.

Lateral colour

This is sometimes referred to as the chromatic difference in magnification because the image heights of an off-axis object vary with wavelength. For the simple positive lens, shown in Figure 1.29, the blue chief ray is bent towards the optical axis more than either the green or the red. The amount of lateral colour in a lens is usually expressed

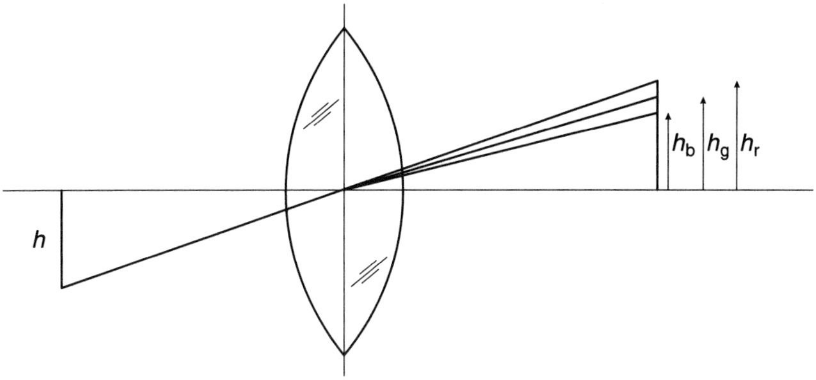

Figure 1.29 Lateral colour

as the difference between the image heights at the extremes of wavelength, divided by the height at the nominal colour. For a white light system at the primary colours,

$$\text{lateral colour} = (h_{red} - h_{blue})/h_{green} \tag{1.34}$$

In practical terms this produces dispersion at the edge of the image and is often difficult to correct. The amount of lateral colour that can be tolerated depends on the use of the lens; but, as a guideline, lenses for liquid-crystal display (LCD) projection should exhibit no more than 1.0 picture element (pel) of colour misregistration at the edge of the projected picture.

Polarization

It should already be clear that the properties of light can be explained by treating light as either a particulate or a wavelike phenomenon. Any electromagnetic wave can be resolved into two components or vectors, which usually travel at the same velocity, at 90° to each other. Ordinarily, light is unpolarized, with equal amounts of energy in the two polarizations, but under certain circumstances randomly polarized light can become plane or linearly polarized. This is done inside a laser cavity by arranging for the light to strike a suitable window at the Brewster angle so that the light is polarized at source (see Chapter 2 on Coatings). More often, randomly polarized light must be converted to a single polarization by using a plane polarizer such as polaroid sheet. By definition, this always throws away at least 50 per cent of the input light; more typically, transmission is even worse, varying from 30 per cent to 45 per cent. If a beam of light is viewed after passage through two plane polarizers then the net amount allowed to pass through the combination depends on the relative orientation of their axes (θ):

$$P = P_{max} \cos^2 \theta + P_{min} \sin^2 \theta \tag{1.35}$$

Here, P_{max} is the maximum transmission of a polarizer and P_{min} is the minimum. If the polarizers are crossed then $\theta = 90°$ and $P = P_{min}$.

Polarization losses are one of the unavoidable loss mechanisms that make liquid-crystal display (LCD) panels so inefficient. An LCD contains a switchable retarding layer sandwiched between a pair of crossed plane polarizers. If the incident illumination is randomly polarized then the first plane polarizer rejects at least 50 per cent of the light. When the display is switched 'on', the retarder rotates the polarization state of the light passing through the first polarizer and aligns it with the second, so allowing transmission.

Many optical materials are isotropic, so that light travels through them at the same velocity irrespective of orientation. Anisotropic materials such as calcite have birefringence, which means that the light in orthogonal polarizations travels at different velocities in the crystal. Since the velocities are different, the refractive index is different, and this causes the two polarizations to get out of phase with each other. This property is utilized in polarization beamsplitters to separate a beam without having to coat any surfaces. (See Chapter 3 on Polarizing beamsplitters.)

In such a crystal, the light that obeys Snell's law at the crystal surface is called the 'ordinary' ray. Not surprisingly, the ray that does not always obey Snell's law on refraction at the crystal surface is the 'extraordinary' ray!

The anisotropic nature of some materials can also be used to change the relative phase of the ordinary and extraordinary rays. A quarter-wave plate introduces a quarter-cycle phase retardation between the two equal energy polarizations, and produces circularly polarized light. If the retardation is anything other than $\frac{1}{4}n$, where n is an integer, and the energy in the two directions is unequal, then the result is elliptically polarized.

CHAPTER 2

Lenses and coatings

Introduction

In this chapter we discuss where and when to use the singlet lens and where and when to apply a lens of more complex form. This discussion ranges over the performance limitations of a single lens and how performance can be improved through the application of more elements, more glass types and different component configurations.

From the simple singlets and doublets, the discussion moves on through the not-so-simple triplet and finally embraces the more complex forms of telephoto lenses, wide-angle lenses, microscope objectives, eyepieces, aspheric lenses, Fresnel lenses and mirror lenses.

Optical coatings and filters are introduced as a means by which better transmission, better contrast and wavelength selection can be achieved. This section also includes a description of hot and cold mirrors, essential for the control of thermal effects for optical systems such as high-power projectors.

The singlet lens

The singlet lens is the simplest optical element, except perhaps the window. It comprises one piece of glass only and is not necessarily curved on both sides. The six fundamental forms of the singlet are presented in Chapter 1 (see Figure 1.6) and comprise the following:

- Biconvex
- Biconcave
- Planoconvex
- Planoconcave
- Meniscus convex
- Meniscus concave

Uses for the singlet lens are many and varied, but its application must be considered carefully because performance can be poor compared with that of more complex lenses. However, in the interest of low cost and low weight it is advantageous to accept the simplest solution that meets the specification, and that solution could be the singlet lens.

The singlet lens is made from one glass type only. Glass exhibits dispersion, the change of refractive index with wavelength, and hence it is seen immediately that a singlet lens will suffer from poor colour performance. For a positive singlet made from high-dispersion glass, the red focus will be formed beyond the blue focus because the red light passes through a lens of lower optical power (i.e. longer focal length). If your application is monochromatic, say a laser collimator, then this poor colour performance does not concern you; or, perhaps, your application is radiometric or photometric only (see Chapter 6), and imaging performance with colour is not so important.

A balance has to be struck between acceptable imaging performance and cost. The singlet lens will perform adequately for some applications and not at all well for others. How do you decide when the performance of the singlet is adequate to achieve the purpose of the optical system? You must know what you want from the optical system and be able to specify it in a sensible way.

For example, say you are designing a position sensor that uses a linear CCD array as the detector and that each element of the array is on a pitch of 10 μm. You might decide that in the interests of high resolution the maximum permissible spot diameter produced by the optical system is 10 μm. Can you use a simple biconvex singlet lens?

If the lens is diffraction-limited then, at best focus, more than 80 per cent of the optical power will be contained within a disc of radius:

$$r = \frac{1.22\lambda f}{D}$$

where λ = wavelength
f = lens focal length
D = lens diameter.

This is known as the Airy disc formula.

Consider two well-separated but visible wavelengths, one a red at 706.5 nm and the other a blue at 404.7 nm used with a 50 mm-focal-length lens of 40-mm diameter. The two diffraction-limited spot diameters are given by:

$$d\,(\text{blue}) = 2.44 \times 404.7 \times 10^{-9} \times 50 \times 10^{-3}/40 \times 10^{-3} = 1.23\,\mu\text{m}$$
$$d\,(\text{red}) = 2.44 \times 706.5 \times 10^{-9} \times 50 \times 10^{-3}/40 \times 10^{-3} = 2.16\,\mu\text{m}$$

Note that resolution performance is better at the shorter wavelength.

These spot diameters are well within our performance specification, but what effect does the shift in focal position with wavelength have?

Let us say that the biconvex singlet has surfaces with a radius of curvature of 50 mm and is manufactured from Schott's BK7 glass, which has a refractive index of 1.5129 at 706.5 nm and 1.5302 at 404.7 nm. Focal length is given (by the thin-lens formula) as follows:

$$1/f = (n-1)(1/R_1 - 1/R_2)$$

which for a biconvex lens, $R_1 = R_2$, simplifies to

$$1/f = (n-1)2/R$$

and for each of the wavelengths we obtain a focal length of 48.7 mm for the red and 47.1 mm for the blue.

Note that the blue focal length is shorter than the red.

Now, with reference to Figure 2.1, the two focal planes are separated by 1.595 mm. If the focal plane of the CCD array is placed at the blue focus then the red blur spot will be of radius 0.654 mm (by similar triangles), which is clearly outside our 10-μm diameter requirement. A best-focus spot between the two focal positions would provide a balance, but even so the blur spot diameter will only approximately halve, giving diameters of about 500 μm, which completely dominates the Airy disc contribution.

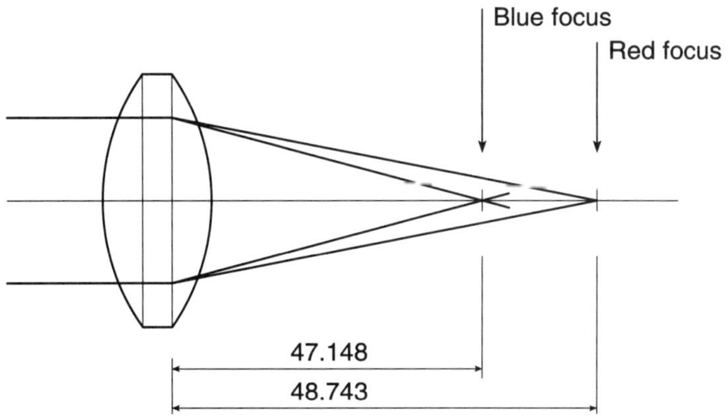

Figure 2.1 Separation of focal plane with colour

Clearly the singlet lens selected will not achieve the specification, *even though the Airy disc diameter was small.* The performance of the system can be improved of course by use of a planoconvex lens (not so fast), by reduction of the lens diameter (large f-number, slower lens), by reduction of the spread of wavelengths (narrow bandwidth) or through use of a achromatic doublet (see next section).

The analysis above assumed a perfectly collimated input, without a field angle. A finite field angle will increase image size and produce curvature of the image plane.

So, to improve colour performance an achromatic doublet is required, and to improve both colour and field performance nothing less than a triplet should be considered. Admittedly, this is a gross generalization, but the trend is accurate!

The best imaging condition for a simple, biconvex lens is obtained at the so-called $4f$ condition. This is achieved when the image conjugate (v) and object conjugates (u) are equal,

$$1/u + 1/v = 1/f$$

and with $u = v$

$$2/u = 1/f$$
$$u = 2f$$

Thus, for a thin lens, a unity magnification image is produced when the object is placed at twice the focal length of the lens from the lens. A distance of $4f$ exists between object and image and hence the $4f$ condition.

In general, singlet lenses are best used either for narrow-colour-bandwidth systems (lasers) or where colour performance is not important.

For collimation applications a planoconvex form is best with the plano side towards the short conjugate (focus). Best performance is achieved with the steepest side of a singlet toward the longer conjugate.

The singlet lens: guidelines

By way of conclusion to the singlet lens some basic guidelines are offered as follows. Please recall that these are rules of thumb only. Detailed analysis of the application will be required to identify the correct optical solution. These rules will put you in the right ballpark, but will not provide a functional design solution unless the application is very simple indeed.

Specific guidelines

1. The object–image relationship is given by

$$1/u + 1/v = 1/f$$

2. The magnification is given by

$$M = v/u$$

3. For equi-biconvex lenses the best imaging is given by the $4f$ condition (unity magnification).
4. For planoconvex or unequal biconvex lenses the plane or flatter side of the lens should be toward the shorter conjugate for best imaging.

General guidelines

5. Use for narrow-bandwidth applications.
6. Use for small field angles.
7. Use for low-resolution applications.

The doublet lens

The doublet lens is manufactured from two differing, but complimentary, glass types in order to achieve an improved colour performance compared to the singlet lens.

An achromatic doublet is specifically corrected at two wavelengths, that is the focal position is identical for two different colours. This has the effect of reducing secondary spectrum, the magnitude of variation of focal position with colour, near the two colours of interest. This in turn improves polychromatic resolution and MTF performance.

The extra degrees of freedom associated with the doublet when compared with the singlet (more surfaces, thicknesses and materials) make possible much-reduced spherical aberration and coma when compared with a singlet of the same f-number, thus further enhancing resolution performance.

In very general terms the two glass types and shapes usually selected for the design of an achromatic doublet are a positive crown element and a negative flint element; see Figure 2.2.

Glass types, for the visible region, are somewhat arbitrarily divided into crowns and flints. The crown types are usually of low dispersion and low refractive index, while the flints are of high dispersion and high index.

How does this combination of glasses and shapes conspire to produce a compound lens that has identical focal lengths at two different wavelengths? Consider the individual components. The positive crown element is designed as a more powerful lens than the negative flint. However, the higher dispersion characteristic of the weaker flint element can result in both having the *same* physical separation between, say, the red and blue focal positions. A positive singlet lens has the effect of placing the blue focus nearer to the lens, while the converse is true of a negative singlet. Therefore the combination of positive crown and negative flint elements can be used to fully correct focal positions for two independent wavelengths.

This colour balance between lens power and dispersion leaves lens shape as a near-independent variable, which can be used to correct both spherical aberration and coma.

By arranging for the facing curvatures of the two elements to be identical, the lenses can be cemented together to form the very robust cemented achromatic doublet. If the two elements are not cemented then the lens is referred to as an air-spaced doublet. The air-spaced doublet can find very useful application where high temperatures may be experienced and the differing thermal expansions of the two glass types could be problematical.

As with the singlet lenses discussed in the previous section, certain general rules apply to the application of doublet lenses. The steepest lens curvature is usually positioned towards the longest conjugate. In the case of, say, a telescope objective

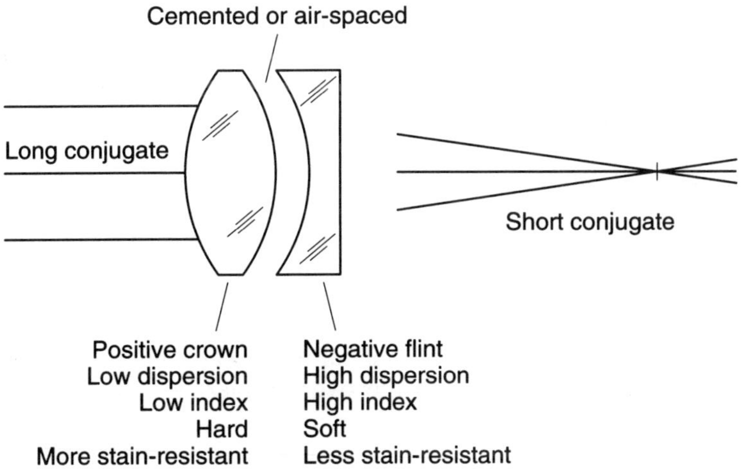

Figure 2.2 Generalized achromatic doublet

this would be the crown element, which is a harder glass than the flint and more stain-resistant. Even more generally, doublets tend to be used for applications of one infinite conjugate such as the telescope objective already mentioned.

Some achromatic doublets are referred to as athermalized, which means that their colour correction is maintained over a specified temperature range. As a glass warms, its dimensions increase, which can result in an increased back focal length (BFL). However, for some glass types the refractive index increases with temperature, making the lens more powerful and thus compensating for the increased BFL.

The doublet lens: guidelines

1. Much-improved bandwidth performance compared with the singlet.
2. Improved resolution performance, even monochromatically, compared with the singlet, due to low spherical aberration and coma.
3. Potential for athermalized designs.
4. Good full-aperture image performance due to low spherical aberration.
5. Good off-axis performance, due to low coma, up to a practical limit of about 2.5 °.

The triplet lens

The triplet lens has three elements and can be found in a large variety of different configurations. The triplet really breaches the artificial boundary established here between the lenses of simple and complex types.

The additional glass element of the triplet introduces more degrees of freedom for the optical designer, and in principle this is just enough to compensate for all of the major aberrations. So lenses of very high performance, diffraction-limited, are possible.

The extra element allows for a variety of improvements over and above the doublet. For example, a triplet with an achromatic, doublet-like front group with an air-spaced final element acting as a field flattener will significantly improve the off-axis – or field – performance of the combination. This is often referred to as the Cooke triplet (or Taylor triplet) configuration and is illustrated in Figure 2.3.

The Cooke triplet serves as an excellent example of a lens in which all primary aberrations are corrected. The lens comprises two positive crown elements that surround a negative flint. Owing to its simplicity, large relative aperture, flat field and low distortion, it is still employed as a photographic objective, despite its having originated some hundred years ago.

One common application of the triplet configuration is as a high-powered projection lens where the air spacing assists with the overall tolerance of the system to high temperatures.

A symmetrical cemented triplet will perform extremely well as a wide-field relay lens. This lens would be best applied in symmetrical applications of unity magnification.

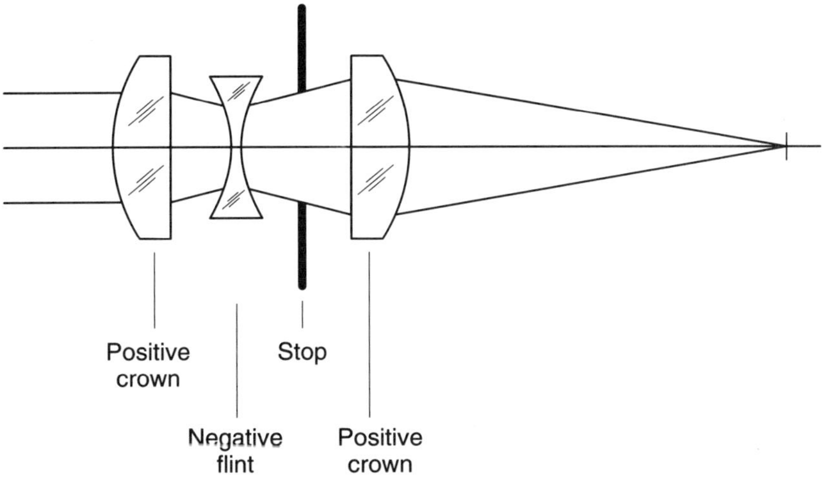

Figure 2.3 *Cooke triplet form*

The third element of the triplet can be selected for its dispersion value, and hence serve to reduce further the secondary spectrum of the doublet. It is possible with this extra freedom to arrange for the lens to be corrected at three wavelengths. The lens is then said to be apochromatic.

An air-spaced triplet lens will have six surfaces, all of which will reflect a proportion of the incident light according to Fresnel reflection losses. As a rule of thumb, an air–glass or glass–air interface, in the visible region of the spectrum, will exhibit an approximate 4 per cent loss. So, with six surfaces a transmission factor due to reflection only can be estimated as

$$0.96^6 = 0.78$$

Thus only 78 per cent of the light is transmitted. The lost 22 per cent can interfere with the desired light path, leading to such unwanted effects as low contrast, stray light and ghost imaging. The simple way to avoid this is to antireflection-coat the lenses. This can be achieved with either a simple broad-band reflection coating, which would reduce the loss to about 1 per cent at each surface ($0.99^6 = 0.94$), or with quality multilayer coatings, which under some circumstances can reach very low reflection values.

The triplet lens: guidelines

1. Much better off-axis performance than a doublet. However, on-axis performance can be similar to the doublet, so field angle will influence your decision between doublet and triplet.
2. Low f-numbers are possible, which makes the triplet very suitable for camera lenses and projectors or any system where light-gathering power is important.
3. More lens surfaces result in the need to control reflection and stray light. Antireflection coatings are required.

4. The triplet can achieve three-wavelength colour correction, resulting in the so-called apochromatic lens.

Complex lenses

This chapter has so far discussed the relevance and application of singlets, doublets and triplets. We now move on to discuss some lenses that have a particular function, and to highlight some issues associated with their design and application.

Telephoto lens

A telephoto lens has the property of having a long focal length, while maintaining a relatively short back-working distance or BFL. This is most readily seen in a photographic lens, where a long focus is achieved with a relatively short lens-to-film distance.

Figure 2.4(a) illustrates how the telephoto arrangement can be achieved. The front element of a lens pair is positive, while the rear is negative. The rays brought toward the focus f(1) by the positive lens are slowed by the negative, and hence the focus is extended to f(2). The BFL is clearly the distance between the vertex of the negative lens and the focus f(2). However, the ray bending to form the focus f(2) appears to come from a plane passing through the point N_2 which is the rear nodal point of the system. For a collimated light input the distance from N_2 to f(2) is the EFL, and the telephoto condition is achieved.

The telephoto lens is constructed rather like a Galilean telescope, but instead of the elements being arranged for afocal operation they are displaced for one finite and one long conjugate.

A wide separation of the two power blocks is required for maximum benefit. The negative lens can be formed from several components, which will help in the control of distortion. Colour correction is achieved by independently achromatizing each of the two groups. Telephoto lenses are usually only applied to relatively small fields, and hence coma and lateral colour aberrations are usually small.

One such shared power scheme is illustrated in Figure 2.4(b).

A collection of lenses that perform a particular function within the complete lens is called a group. Each lens of a group is called an element. Thus, complex lenses are constructed of groups and elements. The telephoto lens illustrated in Figure 2.4(b) is constructed of two groups and five elements.

It is usual for a telephoto lens to have both nodal points in front of the optical system. The ratio x/EFL (see Figure 2.4(a)), is called the telephoto ratio and is less than 1 for a telephoto lens. Alternatively, the distance x can be expressed as $k \times$ EFL, where k is known as the telephoto ratio.

The inverted, or reversed, telephoto lens has the reverse function of the telephoto. This lens finds application where short focal length and long working distance are required. The principle is the reverse of the telephoto, that is, the first element is diverging and the second converging (see Figure 2.5).

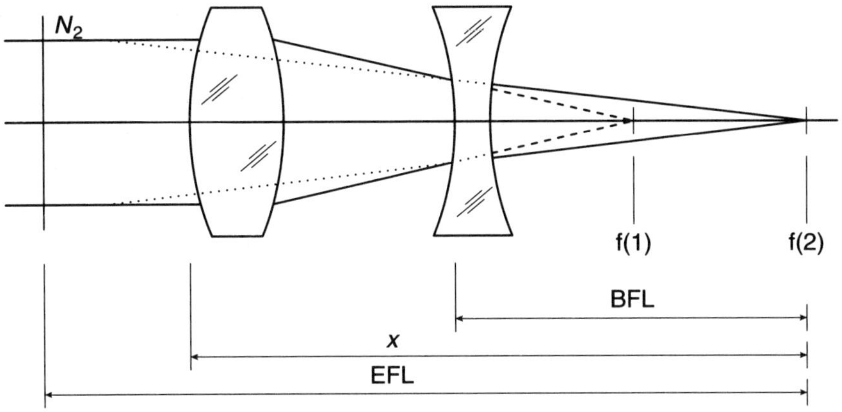

(a)

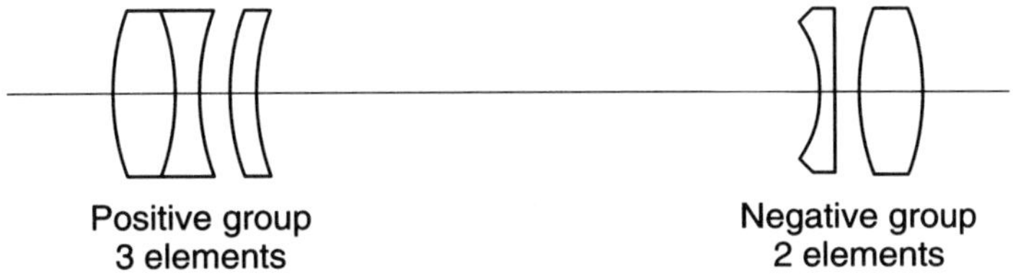

(b)

Figure 2.4 (a) The telephoto lens, (b) Multi-element/group telephoto

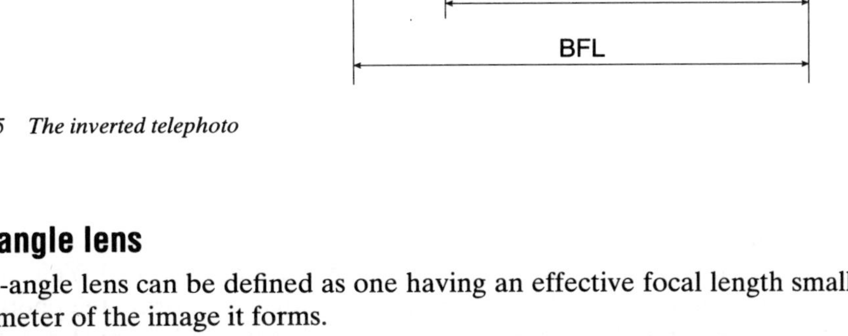

Figure 2.5 The inverted telephoto

Wide-angle lens

A wide-angle lens can be defined as one having an effective focal length smaller than the diameter of the image it forms.

The field of view or coverage for wide-angle lenses is large. The lenses find application in many areas, including wide-field radiometry and photography in confined areas or, simply, where an unusual effect is wanted.

Many wide-angle lenses are of a construction that is symmetrical about the aperture stop. The components about the stop are usually of meniscus shape. This is a consequence of the function of the wide-angle lens, which is to reduce the incidence angle of rays from the extreme of the large field. Indeed, very simple forms of the wide-angle lens have a pair of positive meniscus lenses placed symmetrically about the aperture stop (see Figure 2.6(a)). This form of the lens can provide an acceptable performance for fields of ± 70 ° at aperture ratios of about f22.

To improve performance, an outer, symmetric shell of negative meniscus lenses can slow the ray convergence angles to an extent where an inner pattern of more traditional lenses can take over. This approach enables the inner core of lenses to address aberration correction. Such a layout is suggested in Figure 2.6(b).

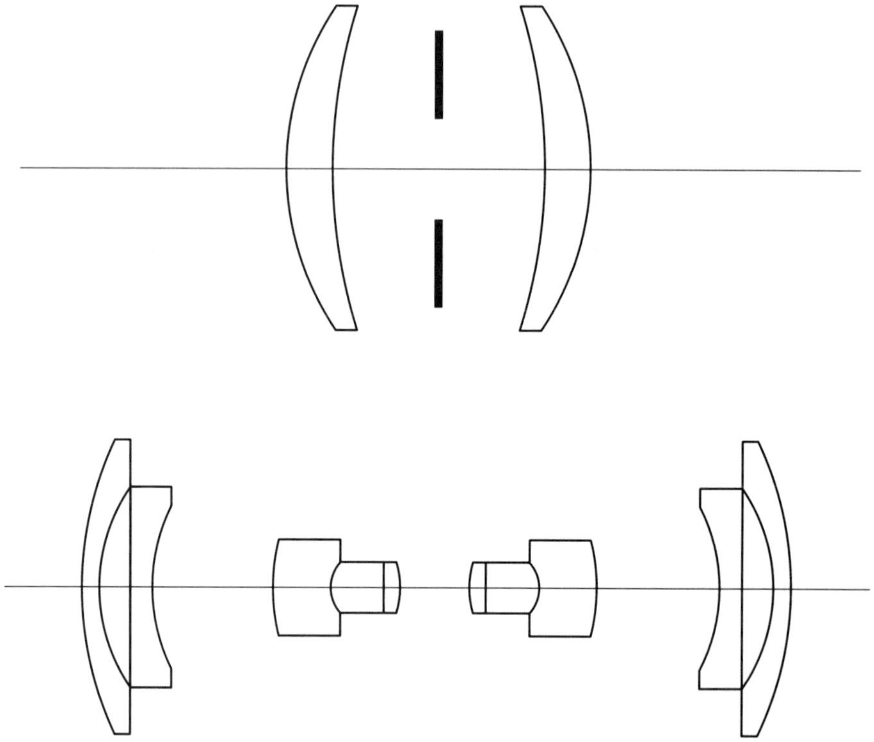

(a)

(b)

Figure 2.6 (a) Basic wide-angle lens. (b) Complex wide-angle lens

To increase further the field of view of a wide-angle lens the symmetrical approach is abandoned and a much-altered form of the reverse telephoto adopted. You will recall from the discussion on the telephoto lens above that a reverse telephoto has a negative front element and a positive rear one such that the EFL is much shorter than the BFL; when this type of lens is taken to practical limits we have what is generally termed a fisheye, or ultra-wide-angle, lens. With such lenses it is not unusual to achieve fields of view of more than 180 °. A typical form is illustrated in Figure 2.7.

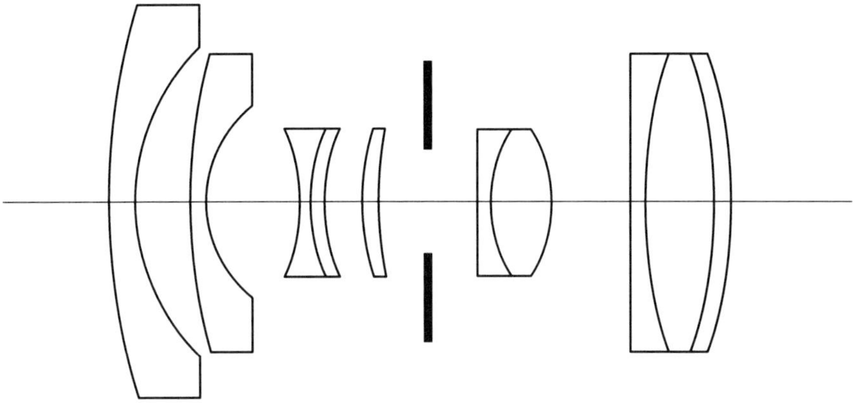

Figure 2.7 Ultra-wide-angle lens

Characteristics of the fisheye lens are extreme distortion (imaging an entire hemisphere onto a flat plane) and mobility of entrance pupil (light bundles diametrically opposed cannot enter the lens through the same piece of glass).

Microscopes

The compound microscope

The compound microscope comprises three main parts: the objective, the eyepiece and the tube or body. The objective provides a magnified image of the object, which is further magnified by the action of the eyepiece and eye together. The body keeps the parts together and excludes the ingress of ambient light.

The parts of the microscope and their relationship with each other are shown in Figure 2.8. The objective forms a magnified image of the object in the focal plane of the eyepiece. This intermediate image is at the field stop of the eyepiece. The eyepiece acts as a simple magnifier and places the image of the object at infinity for relaxed viewing by the eye.

The magnification of a microscope is found by multiplying together the individual magnifications of the objective and the eyepiece. For example, a ×20 objective used with a ×10 eyepiece will provide a magnification of ×200. This relationship only holds true when the microscope components are used at the correct conjugates as described below. The magnification and numerical aperture of the objective and the magnification of the eyepiece are usually engraved on the lens barrels.

The design and form of a microscope and its component parts have to conform to certain basic rules so as to ensure compatibility between different manufacturers. These rules are now summarized and are illustrated in Figure 2.9.

Each microscope objective has a shoulder, 150 mm from which an image of the object will be formed provided that the object is placed at the parfocal distance. This

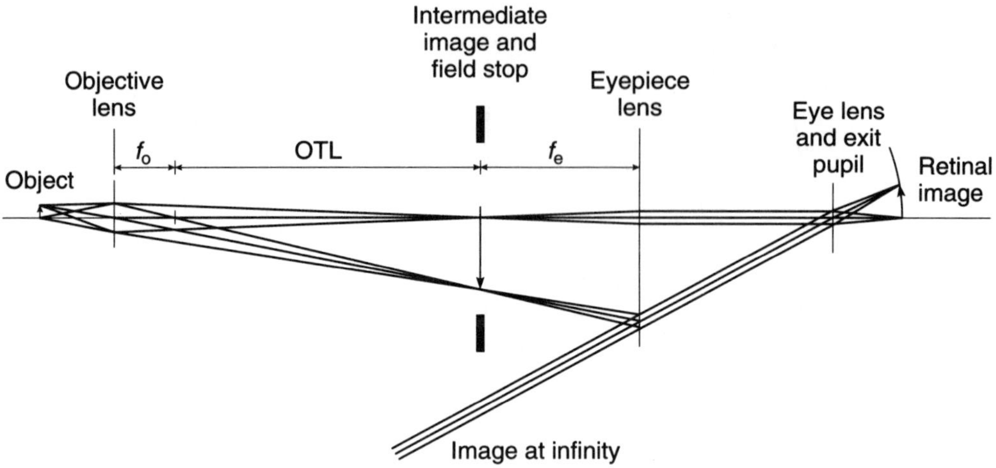

Figure 2.8 Compound microscope. Abbreviations: OTL, optical tube length; f_o focal length (objective); f_e focal length (eyepiece)

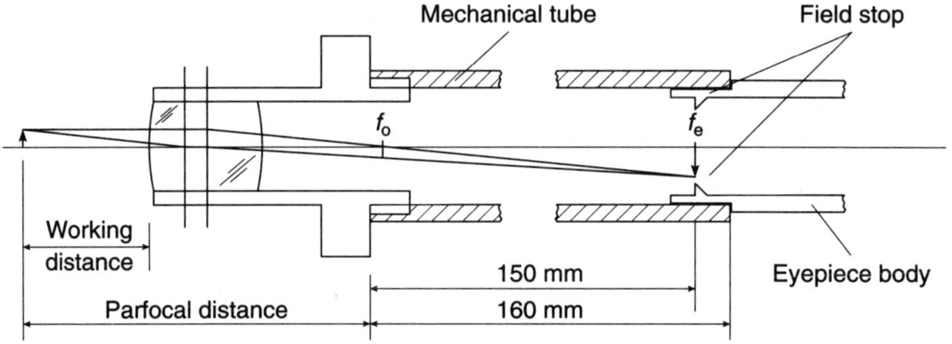

Figure 2.9 Microscope geometry

intermediate image is formed at the field stop of the microscope, which is located 10 mm inside the barrel of a standard 160-mm microscope tube. The rim of the eyepiece abuts the tube, and hence the object plane for the eyepiece is fixed. This mechanical tube length, or intershoulder distance, is precise and fixed at 160 mm.

Setting microscope objectives to their parfocal distance ensures that the lenses are performing with least aberration and therefore giving the best possible image. Standard objectives are designed to throw an image 150 mm from the shoulder when the object is placed at the parfocal distance. Parfocality means that the object-to-image distance is a constant across a range of microscope objectives. This means that parfocal objectives of differing magnifications can be exchanged without significant refocusing. To maintain parfocality, and as a general rule, working distance falls as magnification increases and focal length decreases. Different manufacturers will almost certainly have slightly different parfocal distances for their 'standard' objectives – check before changing manufacturer.

The optical tube length of a microscope is the distance between the back focal point of the objective and the focal plane of the eyepiece. The back focal point of an objective is usually internal to the objective construction.

The mounting thread of a microscope objective is a standard Royal Microscopical Society (RMS) fitting, and is defined according to British Standards Institution (BSI) specifications. Details of this standard thread are given in Table 2.1. Such standardization limits the aperture of a microscope objective to a maximum diameter of about 17 mm, and more practically to about 11 mm.

Table 2.1 The RMS thread

	Male thread	*Female thread*
Pitch	36 threads per inch, Whitworth form	
Major diameter	0.7982 in 0.7952 in	0.8030 in 0.8000 in
Pitch diameter	0.7804 in 0.7774 in	0.7852 in 0.7822 in
Minor diameter	0.7626 in 0.7596 in	0.7674 in 0.7644 in

Resolution, or resolving power, is set by the numerical aperture (NA) of the objective. Lord Rayleigh's criterion is applied as follows:

$$\delta l = 1.22 \times \lambda \times f/D \tag{2.1}$$

where δl = distance separating central maxima of two Airy discs such that the discs are just resolved
λ = wavelength of light
f = focal length of the objective
D = diameter of objective.

Recalling that $f\# = f/D$ for a system with one long conjugate and that $f\# = 1/(2 \times \text{NA})$, Equation (2.1) can be rewritten as

$$\delta l = 1.22\lambda/(2 \times \text{NA}) \tag{2.2}$$

A microscope objective of good construction and design having a numerical aperture of, say, NA = 0.17 and operating at a wavelength of 550 nm can resolve objects separated by

$$\begin{aligned} \delta l &= 1.22 \times 550 \times 10^{-9}/(2 \times 0.17) \\ &= 1.97\,\mu\text{m} \end{aligned}$$

Rounding δl to 2.0 μm and taking the inverse gives the limit of the spatial frequency just resolved by the objective, 500 l/mm or 250 lp/mm. This resolution is only just achieved, which means that the contrast between lines will be poor. However, at half this frequency, 125 lp/mm contrast will be good and, assuming the absence of other image-degrading aberrations, imagery will also be good.

Light gathering ability is also related to numerical aperture. The higher the numerical aperture, the better the light gathering ability of the objective. Typical NAs range from about 0.1 for low-power objectives through to 1.4 for high-power ones. In air, a NA of 1.0 cannot be exceeded, and the highest NAs are only achieved through use of an oil-immersion objective (see below).

Objectives

Microscope objectives fall into three main categories – refracting, reflecting and oil-immersion.

The refracting type is the most commonly encountered, and encompasses a large magnification range from typically ×1 through to as high as ×40 or ×50. Various types are available, including the following:

- Flat mounted.
- Standard parfocal.
- High-precision with very low wavefront error for laser or interferometric applications.
- Interferometric types: literally, small interferometers for the measurement of surface form errors and thin film thickness.

Microscope objectives are available for use outside the visible spectrum, with one common application being the launching of near-infrared laser light into fibre optics.

Reflecting microscope objectives having no refractive components are useful for the complete elimination of chromatic error. These types can be used over a very broad spectral range, but are often larger than the refractive equivalent.

The other objective type to be considered is the oil-immersion. Oil has a higher refractive index than air, and therefore the numerical aperture of the objective can be increased by substituting oil for air between the object and objective. The higher the NA, the better the resolving capability and light-gathering efficiency of the objective. Numerical apertures of greater than one are often achieved.

For example, an objective of, say, 20 mm parfocal distance when set for a tube length of 160 mm will yield a magnification of 160/20 = 8×, which when used with a 10× eyepiece gives a total magnification of 80×. However, an oil-immersion type objective with high NA might have a parfocal distance of only 1.6 mm, which gives 100× for the objective and 1000× for the system.

Eyepieces

Various eyepiece designs are available. Two common types are the Huygens and the Ramsden. The major difference between these two is that the Huygens alone forms the intermediate image inside the eyepiece (remember: Huygenian = H = Hidden). The form of each type is shown in Figure 2.10.

Merits of the Huygens type include low purchase price, relative freedom from lateral colour aberration and a protected image plane for reticules or cross hairs. The biggest disadvantage can be the small eye relief of only a few millimetres.

The Ramsden type will magnify a reticule or cross hair placed at the intermediate

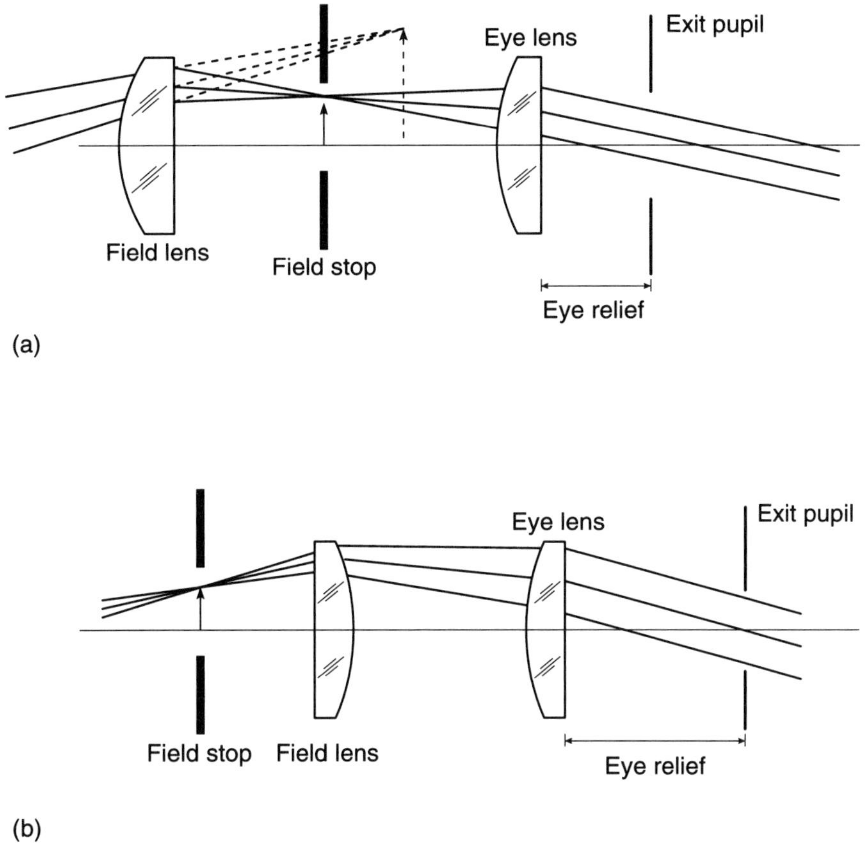

Figure 2.10 Eyepieces: (a) Huygens type, (b) Ramsden type

focus and is easily accessible. It has a longer eye relief, of the order of 12 mm or more. The Kellner-type eyepiece is a development of the Ramsden type, where the eye lens is achromatized, which yields a substantial increase in image quality, but at the expense of eye relief.

Aspheric lenses (aspheric = non-spheric)

An aspheric lens is usually defined mathematically by a function that determines departure from a sphere. That is to say, a base curve – spherical, parabolic, or elliptical – is defined and departure from that curve expressed as a power series.

A typical form, for a rotationally symmetrical surface, is

$$\delta X = a_1 Y^4 + a_2 Y^6 + a_3 Y^8 + a_4 Y^{10}$$

where X is the distance along the X-axis (optical axis) and Y the height from the optical axis, the departure so calculated being summed with the X-value returned from the base curve (see Figure 2.11).

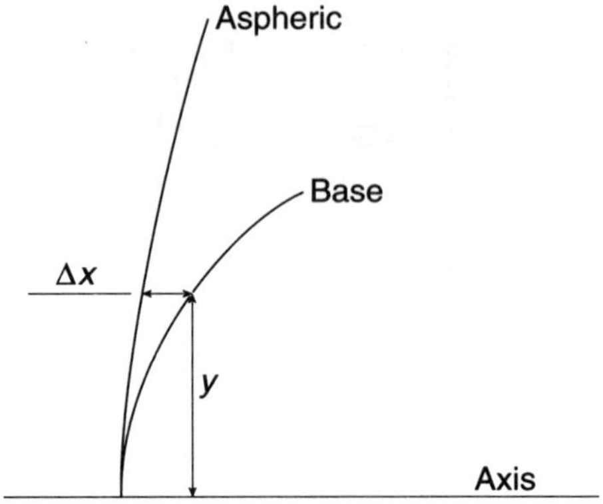

Figure 2.11 Aspheric lens

Many modern optical design programs will accept conic or spherical base surfaces as defined by a conic constant. Conics, so called because they are obtained by the intersection of a plane and a cone, are characterized by the value of an associated constant, the eccentricity (ε).

The conic surface is

1. An ellipse if $\varepsilon<1$.
2. A parabola if $\varepsilon=1$.
3. A hyperbola if $\varepsilon>1$.

The conic constant is defined as follows:

$$k = -\varepsilon^2$$

The aspheric lens is finding increasing use in many varied applications, among them:

Fast condenser

This is employed where it is essential to place a lens as close as possible to a light source to engineer as much light collection as possible. Very fast condensers may have f# as low as f0.5 or f0.6 and still usefully gather light. This type of lens, as it is not used in an imaging application, can usually be of relatively crude construction, for example moulded and fire-polished.

Best-form lenses

Aspheric lenses are particularly suited to applications where a single lens is required to collimate, or work with, a laser. The laser works at a single wavelength only, thus eliminating chromatic errors, and the extra degrees of freedom allowed with an aspheric design means that excellent performance can be obtained from a singlet.

Micro-imaging

Micro-imaging lenses are now used frequently in applications such as secure television. Here a miniature lens might be cemented directly to a CCD imager to produce a tiny but highly capable camera system.

Fresnel lenses

The Fresnel lens can be considered as a crushed form of singlet lens, spheric or aspheric, that has its power divided among many different facets; see Figure 2.12. The lens typically has a circular symmetry and can comprise very fine facets, as with an overhead projector condenser, or coarse facets, as with a theatre spotlight projection lens.

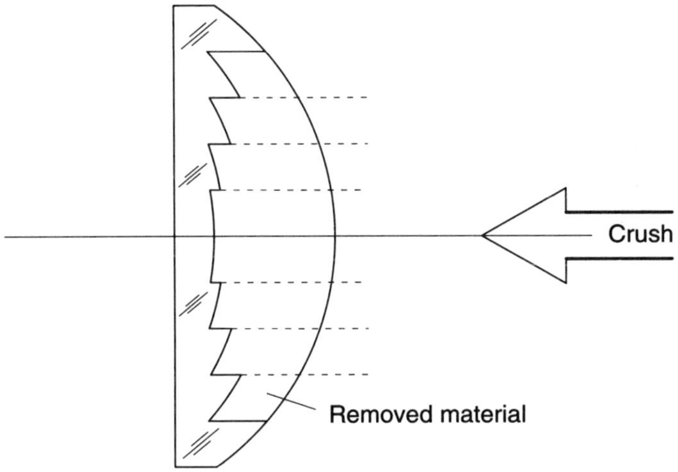

Figure 2.12 Fresnel lens

The crushing analogy is not strictly accurate, as is illustrated in Figure 2.13, where superimposed ray paths are shown. From Figure 2.13 it is clear that the ray traversing the lens before and after crushing will exhibit different characteristics. The ray emerging from 0 traverses the pre-crushed lens and emerges with angle θ_1 to the

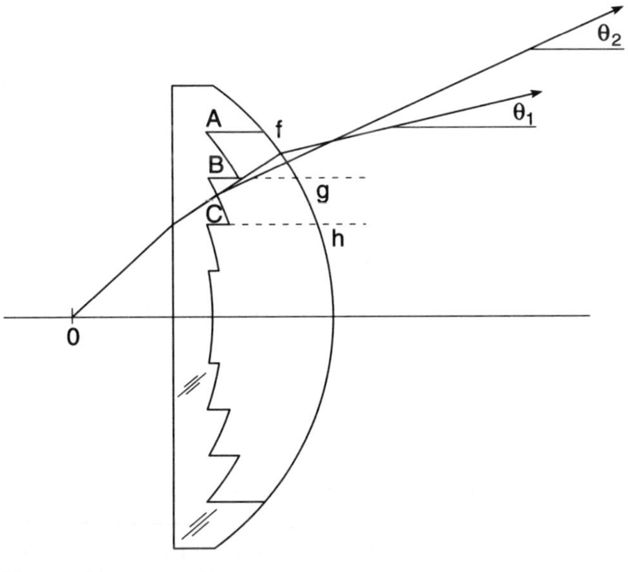

Figure 2.13 Fresnel lens and image quality

optical axis. The same ray, post-crushing, is refracted by facet C, which has less power than facet B by virtue of representing the lens surface g to h, not f to g.

A thin Fresnel lens manufactured with many small linear facets will match closely the original lens that it mimics. This is true of the overhead-projector type of Fresnel lens. The theatre spot, however, which by virtue of its self-supporting diameter maintains a significant thickness, has relatively large facets, and the effect illustrated at Figure 2.13 is significant. Linear approximation to lens curvature over these large, thick facets will yield a poor performance, and the optical designer has to take recourse to designing each individual facet separately, possibly making use of aspheric curvature to maintain a high performance.

The Fresnel lens in all but its simplest linear facet configuration is not defined by any firm mathematical rote. For best performance, each facet must be considered as an individual optical element. The Fresnel lens is a zonal lens array.

Owing to low materials usage, low mass and ease of replication, Fresnel lenses find many applications where considerations of cost outweigh those of performance.

Fresnel lens applications are many, and include among others:

- Theatre spotlighting, where light weight and thermal resistance with good optical performance is required. For this application it is often necessary to texture the lens plano surface to disguise any source imaging that might occur.
- Overhead projector condenser systems, where low cost and good performance are required with high apertures for efficient light collection.
- The traditional and often very large lighthouse lenses, where the weight of a solid lens would have been prohibitive!

A form of Fresnel lens is used in some car lighting or intruder detector systems where rectangular arrays of what are essentially Fresnel lenses are formed for high-aperture systems. Here plastics are used to very good effect. Thin sections are essential for the intruder detector, so as not to attenuate the infrared signal through the plastic lens.

Mirror lenses

Lenses formed from mirror surfaces mainly find application where large bandwidths demand a non-dispersive optical system, such as a very wide-band thermal imaging system. Alternatively, reflecting optics may be the only viable solution for very high optical powers such as those experienced in thermal laser machining.

Mirror lenses fall into three main categories, the first of which is the single-surface reflector, such as a steering mirror, which requires no further discussion here. The two other categories are described below.

Catadioptric lenses

The catadioptric lens is made up of both reflecting and refracting elements. This combination can be used to good effect when the reflecting or mirror surfaces contain most of the power of the system and the refracting element(s) little or no power. This approach can be used to produce images of good quality with small aberration only.

One example of both reflecting and refracting optics used in the same system for control of aberration is the Mangin mirror. This consists of a reflecting surface with a refracting surface in optical contact. If it is used in, say, a reflecting telescope configuration, with the Mangin mirror acting as the secondary mirror, then the first, refracting surface, can be used to correct the spherical aberration of the second, reflecting surface. In this way, simple, spheric surfaces can be used as secondary mirrors. See Figure 2.14.

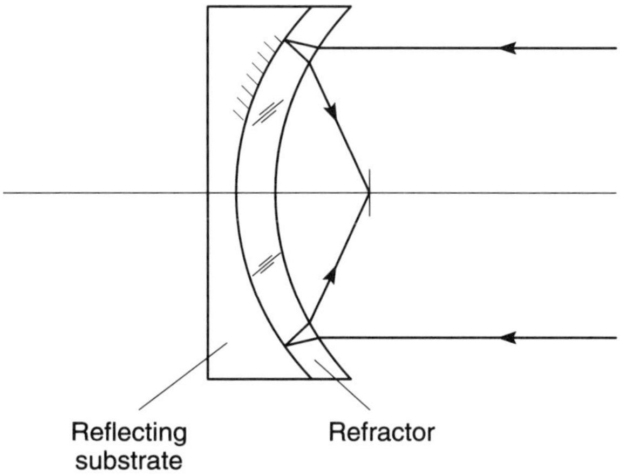

Figure 2.14 Mangin mirror

Catoptric lenses

The catoptric design is truly a 'mirror-only' system, where all the optical power of the system is provided by reflecting surfaces. These systems are applied where good chromatic performance is required over large wavebands or where limited material choice is available.

Mirror-only systems are used extensively where thermal absorption by a refracting element might damage the lens. In these cases the very best in thermal conductivity and optical reflectivity are required. For a high-power CO_2 laser machining centre this might mean delivery optics manufactured from copper, with cooling channels in the rear (air or fluid flow), with gold coatings having a reflectivity of over 99 per cent.

Laser cavities and infrared telescopes are other common applications of catoptric lenses.

Many forms of reflecting telescope have been described and applied. These include the Newtonian and Cassegrainian types as illustrated in Figure 2.15. The Newtonian type uses a parabolic reflector to form a spherical aberration free image of axial objects. A small, flat steering mirror is used to locate this image at the side of the telescope for ease of viewing. The telescope, or more accurately the parabolic reflector, is not free of aberration, and both coma and astigmatism are present for off-axis imaging. However, the degree of astigmatism can be controlled through location of the stop position at or near the focus.

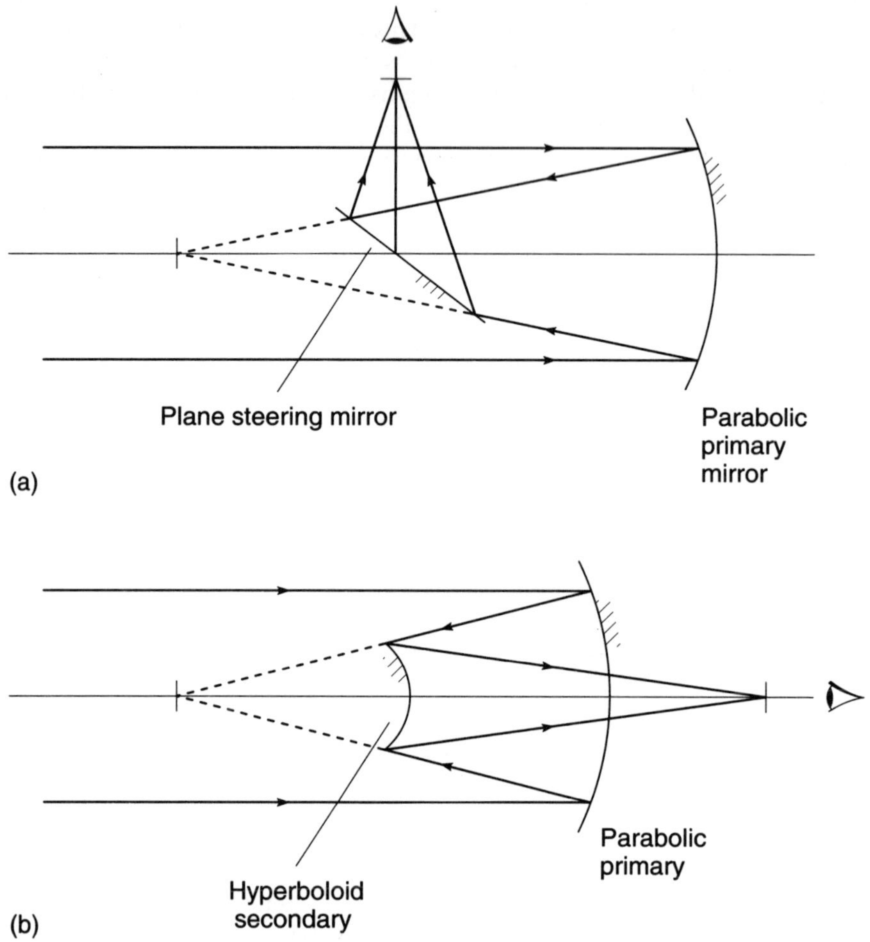

Figure 2.15 Catoptric telescopes: (a) Newtonian, (b) Cassegrainian

The Cassegrainian type is similar in some respects to the Newtonian telescope, but differs in that the secondary mirror is of hyperboloid form that directs light back through a hole in the primary. When the image is at the focus of the hyperboloid the image has no spherical aberration.

Coatings and filters

This section deals with the various types of optical coatings and filters which can be utilized in the design of optical systems. A brief description is included of each coating type, its typical function and the limitations on its use.

The most common form of optical coating is the antireflection or AR coating. To understand why such a layer is required, it is necessary to understand how reflections arise from a boundary between two materials.

Reflection at a material boundary

In Chapter 1, refractive index was described in the context of bending light at the boundary between different materials. A less welcome effect, arising from the difference in refractive indices across a boundary, is that light is also lost by reflection. This effect is often referred to as the Fresnel loss, and apart from reducing the transmission of an optical system it also gives rise to multiple reflections, stray light and veiling glare. Although the contribution to these effects from any single element is seldom large enough to worry about, unfortunately each element contributes a little and the sum can often be disastrous.

For simple cases, where the light strikes the material interface at or near normal incidence, the reflected intensity can be calculated using

$$(1-n_1/n_2)^2/(1+n_1/n_2)^2$$

For an air–glass boundary this reduces to

$$(n-1/n+1)^2$$

If the glass is a common crown, such as BK7, with an index of about 1.5, then the reflected intensity is about 4 per cent. This figure is widely quoted as the 'Fresnel loss', that is 4 per cent loss per interface or 8 per cent loss per lens. For example, an uncoated air-spaced crown triplet will lose about 22 per cent of the original input power to the lens purely from reflection losses. This makes active measures for reducing Fresnel loss very attractive.

Reflected light can also cause problems by reducing the contrast of the image. This is particularly true of projection systems, where a contrast of 100:1 may be needed and uncoated optics are impractical.

The reflectivity depends on both the polarization state (s or p) and the angle of incidence; it can be calculated from:

$$R_p = [\tan(\theta_1 - \theta_2)/\tan(\theta_1 + \theta_2)]^2$$
$$R_s = [\sin(\theta_1 - \theta_2)/\sin(\theta_1 + \theta_2)]^2$$

where θ_1 is the angle of incidence and θ_2 is the angle of refraction.

These two equations allow us to calculate the individual reflectivity of the orthogonal polarization states, s and p. See Appendix A, 'polarization'.

Figure 2.16 shows both s and p reflectivity for light moving from a low index to a high index, i.e. from air to glass. The combined reflectivity for a randomly polarized light source, i.e. one with equal average energy in each polarization state, is simply the average of R_s and R_p.

Again taking the example of BK7, several key points are worth highlighting from this formula as follows.

From low to high refractive index:

1. As the angle of incidence increases beyond 40° the average reflected intensity begins to rise above 5 per cent. So it is advisable to keep angles of incidence below about 30° for uncoated optics.

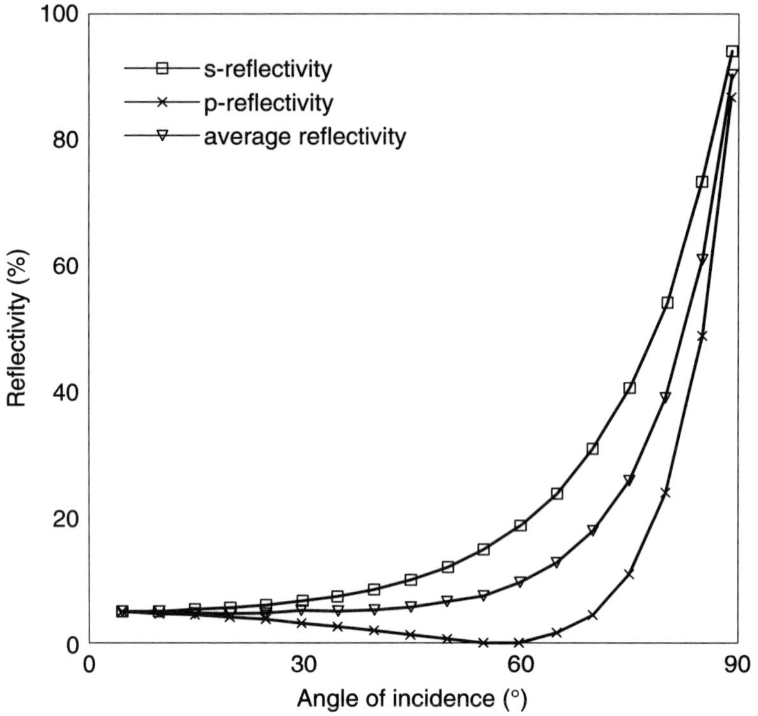

Figure 2.16 Variation of reflectivity with angle of incidence

2. At about 56.6 °, the p-state reflectivity becomes zero while the s component is about 15 per cent. This angle is called the Brewster angle and it can be used to provide linearly polarized light. The Brewster angle for any glass can be found from

$$\theta_b = \arctan(n_2/n_1)$$

3. Beyond 60 °, both s and p reflectivities begin to rise rapidly towards 100 per cent at grazing incidence.
4. The phase of reflected light is shifted by 180 °.

From high to low refractive index:

1. The Brewster angle is reduced by the ratio of the refractive index, i.e. it becomes 33.33 ° for BK7.
2. The onset of total internal reflection occurs at the critical angle, given by

$$\theta_c = \arcsin(n_2/n_1)$$

   which for a BK7–air boundary is about 41.14 °.

3. The phase of reflected light is shifted only if total internal reflection occurs.

Anti-reflection (AR) coatings

For the reasons already outlined, AR coatings are very important, especially in complex optical designs.

An AR coating is analogous to a quarter-wave plate in front of an acoustic transducer or an impedance-matching coil in an electronic circuit. All three use the principle of interference to minimize reflection of the wavefront at a boundary layer.

AR coatings are very thin layers of a material deposited on a substrate (lens, fibre window etc.). Coating thickness is designed so that reflections from the outer face of the coating and from the substrate, are in antiphase and of equal intensity. The result is that they destructively interfere and, ideally, cancel to zero.

Returning to BK7 as an example, for the destructive interference condition to arise, the refractive index of the air, glass and coating material must satisfy the following condition:

$$n_{air}/n_{AR} = n_{AR}/n_{BK7}$$

or

$$n_{AR}^2 = 1.52$$

A suitable material with an index of 1.23 has eluded the coating manufacturers, so, instead, magnesium fluoride (MgF_2) is often used for simple coatings.

Single layer

The most common form of AR coating is single-layer MgF_2. This is largely because its refractive index, 1.38, in the middle of the visible spectrum at 550 nm, is a good compromise for most crown glasses. It is also relatively insensitive to incidence angle and wavelength, remaining below 3 per cent reflection from 400 to 700 nm and from 0 ° to 45 °. A characteristic example is shown in Figure 2.17.

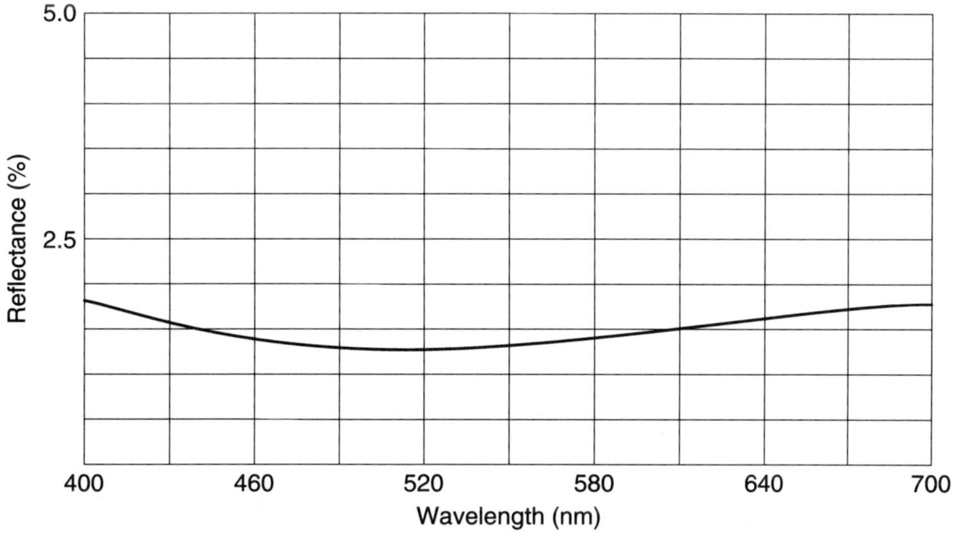

Figure 2.17 Typical antireflection coating

A typical single-layer coating will reduce reflected light to between 1.5 per cent and 2.5 per cent, which would reduce the reflection loss of the previous triplet example to between 9 per cent and 14 per cent. Such a reduction in loss is often sufficient for the designers of many simple optical systems, who neither need nor can afford the cost of more complex coatings.

It is also worth remembering that the efficiency of MgF_2 coating improves as the refractive index of the glass substrate increases, so for some applications it may be better to specify a high-index flint, rather than a crown glass.

Broad-band AR

If the number of coating layers is increased, and the materials in each layer are alternated, then a multilayer coating can be produced which gives low reflectivity, <1 per cent, over typically a 300 nm range. The simplest form of multilayer coating is the two-layer design. The first layer is low-index and the second is a higher-index material. Each layer has an optical thickness of a quarter-wavelength, so both layers introduce a 180° phase shift at the air interface. Destructive interference occurs and the reflection is minimized.

More generally, multilayer coatings include layers with an optical thickness of a half-wavelength at one particular wavelength. To light of that wavelength, the layer is effectively transparent and has no effect, but to other wavelengths it introduces phase changes. This causes varying amounts of interference and reduction of reflection across larger bandwidths than a single layer.

The exact mechanics of multilayer coatings are complex and beyond the scope of this book. However, it is not necessary to have more than a basic knowledge of their operation when it comes to specifying a coating to a manufacturer.

First, decide what wavelength the coating should be optimized for, and what optical bandwidth is required. If the application is a visual one, then the importance of green light overshadows most other considerations, so the peak performance should be specified at 555 nm and the bandwidth should be 400 to 700 nm. If the application is radiometric remember that the product of the source and detector response govern the spectral weighting. For example, a xenon arc lamp has at least 50 per cent of its output in the infrared; if the detector is silicon, then its peak response will be about 900 nm. So although the lamp will produce over 40 per cent of its output in the visible spectrum, the power received will benefit from a coating optimized in the IR. Don't forget that the angle of incidence is also a variable, and that manufacturers can optimize a coating for a particular incidence. This is particularly useful for high-numerical-aperture systems, prisms and off-axis optics.

V-coatings

An extension of the multilayer coating is the narrow-band AR coating or V-coat, so named because of its shape (Figure 2.18). These coatings are optimized for a specific wavelength and angle of incidence. The result is that they can reduce the reflectance to as low as 0.1 per cent over a narrow waveband. Such coatings,

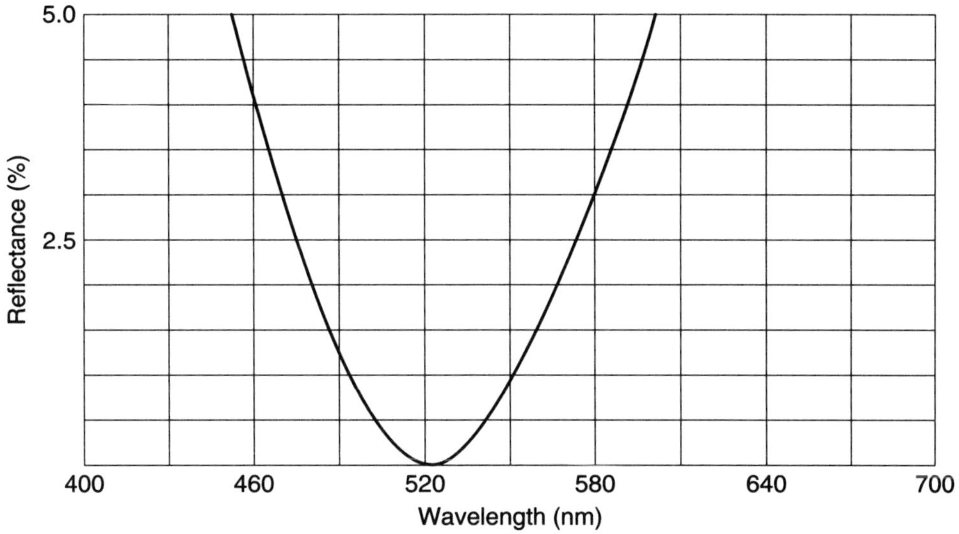

Figure 2.18 Typical antireflection V-coat

although more expensive than a single layer, are particularly useful with narrow-band light sources such as lasers. Many standard coatings can be specified for the common laser types such as argon–ion, helium–neon and laser diodes. Telecommunications wavelengths such as 1300 nm and 1550 nm are also commercially available.

Multilayer filters

Antireflection is not the only application for multilayer interference coatings. A natural extension of the AR V-coat is to selectively transmit only a narrow bandwidth of light, thus creating a narrow-band filter. This is usually done with an interference filter, consisting of a number of stacks or cavities. These behave like mirrors in a Fabry–Perot etalon (see Chapter 4), selectively transmitting a very narrow bandwidth.

A typical example of this type of filter may have a 50 per cent passband from 7 to 10 nm wide and a peak transmission of 40 to 50 per cent. Blocking outside the passband can be as high as 10^{-5}. Filter bandwidth depends on the number of cavities; the more there are, the narrower the pass-band is.

Interference filters often find application in laser systems for the rejection of ambient light or in fluorescence measurements where the stimulation and emitted light are relatively close in wavelength. The spectrum of a typical narrow-band filter is shown in Figure 2.19, along with the important parameters for specifying the filter to the manufacturer.

All multilayer filters are affected by the incident angle of the light, which effectively determines the length of the optical path through the coating. Generally, the peak of

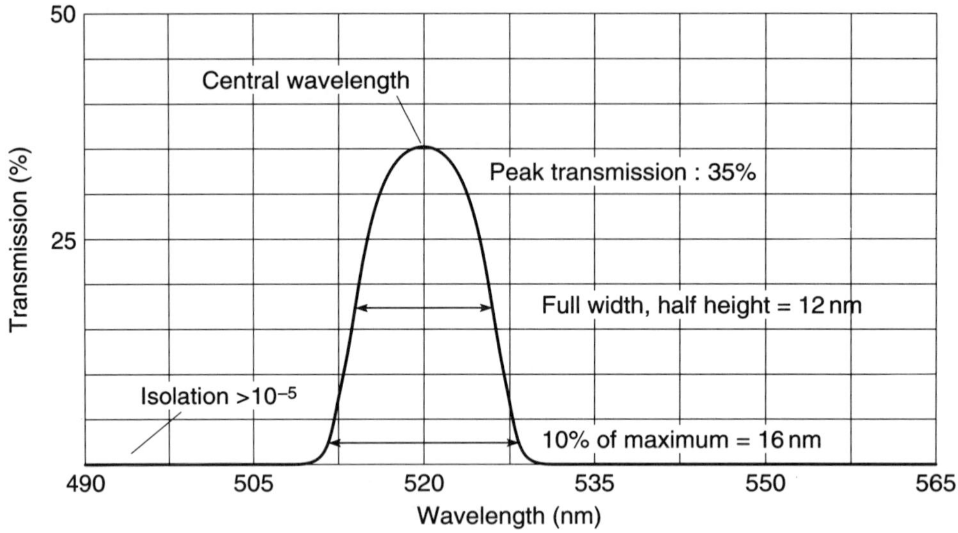

Figure 2.19 Interference filter

the filter response moves to a lower wavelength. This shift can be approximated from the formula:

$$\lambda_\theta = \lambda(n^2 - n_0 \sin^2\theta)^{1/2}/n$$

were λ is the nominal peak of the filter and λ_θ is the peak at an angle θ. In this case, n is the effective index for the whole coating, usually between 1.5 and 1.9, and n_0 is the lower refractive index, i.e. 1 for air. Most manufacturers quote the effective index for standard coatings. But be warned: this formula is seldom useful beyond 30°, when the peak transmission begins to change too.

The angular dependence of multilayer filters can be useful; for example, the filter can be rotated so that it scans or tunes to a particular wavelength. More often it is a drawback and imposes a limitation on the numerical aperture of the incident light.

Dichroic coatings

Interference filters can be designed to select a particular colour or range of colours. A blue dichroic is usually a simple low-pass filter with its 50 per cent transmission point centred at between 470 and 490 nm. A red dichroic is a high-pass filter with the 50 per cent point at between 600 and 620 nm. These two types of filter have only one rising or falling edge, so the transmission throughout the pass band can be high, usually above 90 per cent. This is a significant advantage over gelatine or glass filters, which have a lower transmittance.

Dichroic filters can be designed with very rapid transition edges, going from 90 per cent to <5 per cent transmission in 10 nm. This is particularly useful if colour detection or identification is necessary. Dichroic filters are also more stablc and durable than coloured glasses or gelatin filters. On a suitable high-temperature substrate, thermal shock can be as high as 400 °C for a filter 1 mm thick.

Blocking or isolation outside the bandwidth of interest is usually better than 1 per cent, but with some filters transmittance rises again outside the visible spectrum. This is especially true of blue and green filters, which pass significant amounts of IR, as can be seen in Figures 2.20 and 2.21.

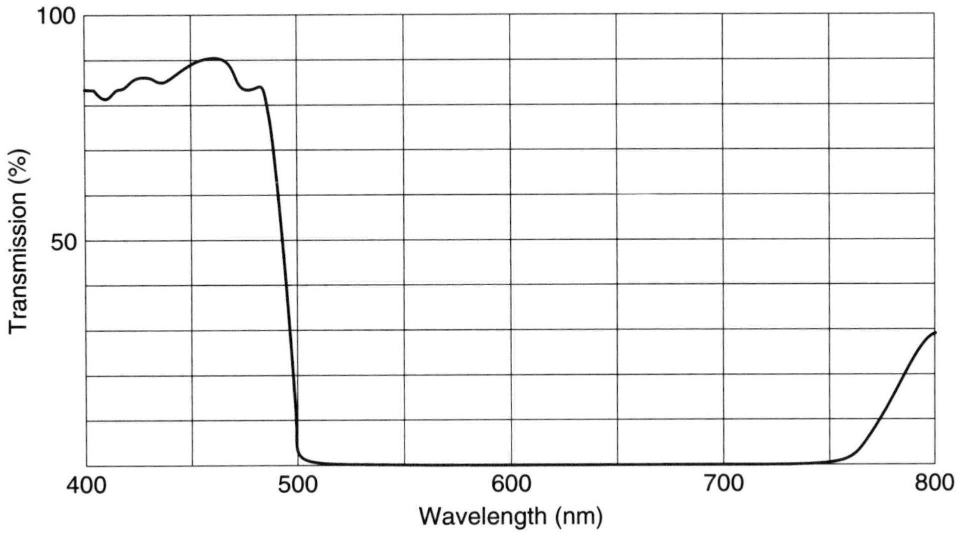

Figure 2.20 Transmission vs. wavelength: blue dichroic filter

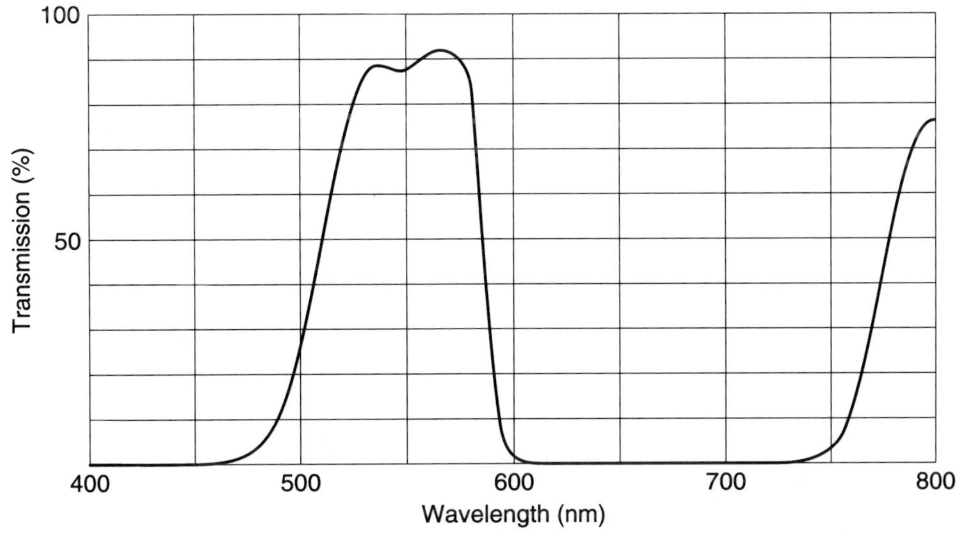

Figure 2.21 Transmission vs. wavelength: green dichroic filter

Bandpass filters, such as a green filter, are possible, but they are generally more complex and expensive because of the need to control two edges. It is usually more efficient to try and separate or create colours using blue and red primaries or the secondary colours, cyan, magenta and yellow. An example of this approach is to be found in a television camera, where the blue and then the red are split off sequentially, leaving the green by subtraction.

Once again, incidence angle has a profound effect on the dichroic performance. In this case it is most often seen as a variation in colour across the field of view. Obviously projection applications, where the eye is very sensitive to colour changes, would be severely compromised by this effect. The solution is simply to ensure that numerical apertures are kept as small as possible; ideally the light should be collimated through the filters.

Some dichroic filters exhibit a thermal drift characteristic, which arises from their porous composition. The coating materials can adsorb water, which is then driven off by heating, and this causes a change in the spectral performance of the filter. Temperature dependence of the filter peak or edge can be as high as 0.133 nm/°C from 20 up to 120 °C. The movement is nearly always towards the lower wavelength, and more significant in the red than the blue. This effect may become a problem when you are specifying a filter for use with a high-powered light source. Ionically, deposited coatings are much less susceptible to this problem but are more expensive. The alternative is to remove the heat before it reaches the dichroic by using a hot or cold mirror. By contrast, IR filters behave in a more linear manner, increasing temperature increases the peak wavelength and vice versa.

Manufacturing tolerances do place a constraint on the specification and cost of filters, but generally a peak or edge position can be held within the range 10 to 15 nm without cost penalties or low yield. Typical examples of dichroic blue, green and red are shown in Figures 2.20 to 2.22.

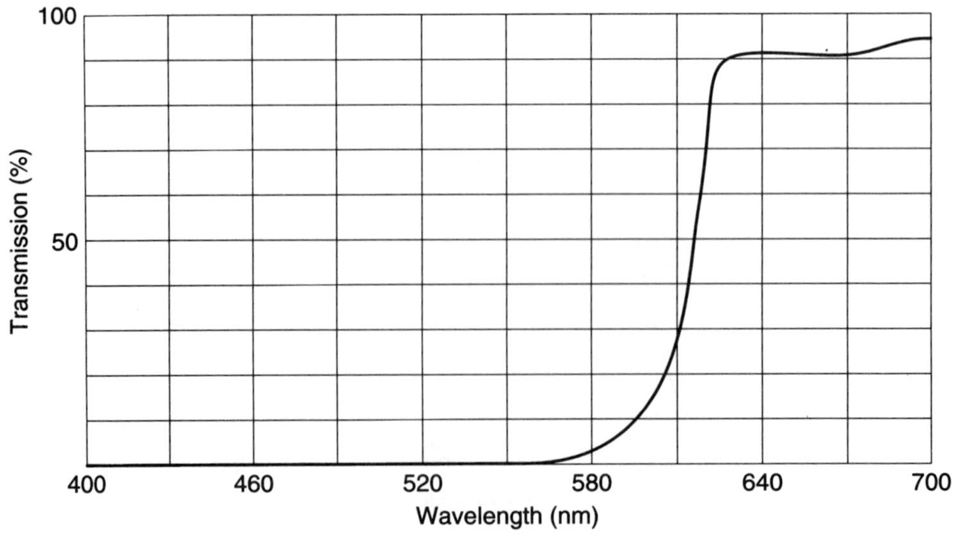

Figure 2.22 Transmission vs. wavelength: red dichroic filter

Hot and cold mirror coatings

These types of coating are really mirror images of each other. A cold mirror reflects visible light and transmits the infrared; a hot mirror transmits the visible and reflects the infrared.

Obviously these mirrors are designed to remove unwanted heat from an optical system; in general, they are more efficient and robust at this task than heat-absorbing glasses. Typically visible-light transmission or reflection is >90 per cent from 450 to 650 nm and the rejection in the IR is 90 per cent or better. However, there are limitations, and the reader should be wary of expecting the complete removal of problematic heat on every occasion.

This is especially true of high-power arc sources such as metal halide lamps, which generate considerable amounts of power in both the visible and the IR spectra. For example, removing 90 per cent of the IR with a cold mirror may not reduce the overall power levels enough to protect an LCD panel from 'bleaching'. Two or more mirrors may be necessary to reduce the heat to acceptable levels, in which process the red part of the visible spectrum will be curtailed. This reduction in the red content can alter colour rendition or white balance and must be guarded against.

Standard cold mirrors can be specified for curved surfaces such as the inside of reflectors or at a 45° angle of incidence. The 45° cold mirror coating is very useful, because it can be used to fold the optical path and save space. However, the 45° angle of incidence leads to a difference in the reflectivity of the s and p polarizations. Since the light is usually randomly polarized, the average reflectivity shown is a characteristic 'knee' in the cutoff edge of the filter (see Figure 2.23).

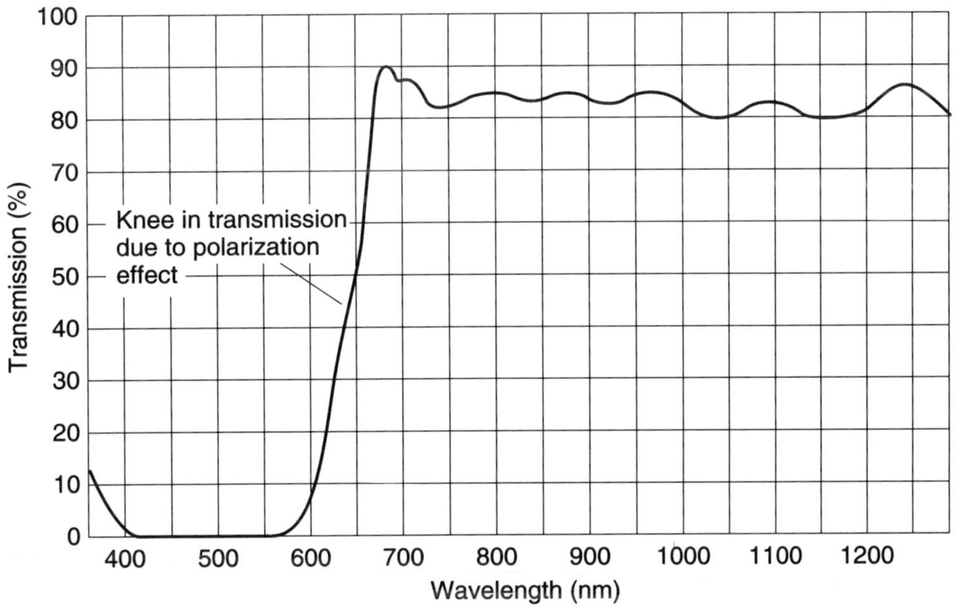

Figure 2.23 Cold mirror transmission spectrum

Gradient index (GRIN) coatings

GRIN coatings are a recent development in filter design, and allow the possibility of extremely sharp cut-filters.

In standard filter coatings the different materials are laid down sequentially, forming a step change in refractive index. In a GRIN coating the refractive index is varied sinusoidally, providing much greater control of performance.

Although they have recently found favour in the specialist design applications such as head-up displays, these coatings are not widely available and are consequently a costly option.

Specifying a coating

If you find yourself in the position of having to commission a custom filter, or you are seeking a standard coating, it is advisable to have considered the following parameters:

- Reflection or transmission – peak value.
- Peak wavelength.
- Bandwidth, full-width, half maximum and 10 per cent width.
- Spectral isolation or blocking required.
- Range and nominal angle of incidence.
- Substrate material.
- Physical size.
- Clear aperture – remember that the coater needs space to hold the filter in the chamber.
- Temperature range.
- Environmental conditions – adhesion, durability, humidity, etc.

Your conclusions will serve as a useful engineering specification and assist the design process.

CHAPTER 3

Prisms and diffraction gratings

Introduction

This chapter discusses some of the wide variety of prisms and diffraction gratings available to the optical engineer, together with their application to modern optics.

Prisms can be employed for many purposes, and find use across a wide range of applications. These applications may embrace a simple beam-steering component such as a robust and stable 45° mirror or the establishment of optical beams with different polarization states. Beam geometry and image orientation can be controlled through the careful application of a few different types of prism. A careful choice of materials extends the useful spectrum of prisms from the deep ultraviolet all the way through to thermal imaging wavelengths.

Practically everyone is familiar with the picture of Sir Isaac Newton using a prism to split light into its constituent colours. This particular optical component, the dispersing prism, finds use in spectroscopic analysis.

Diffraction gratings, both the transmitting and reflecting variety, are presented, together with an indication of some of the cost-versus-manufacturing technique trade offs that might be encountered. Holography and diffraction patterns are now commonplace and are at their most visible with compact-disk technology and security holograms. Even these, the most complex of diffraction patterns, can be considered in terms of basic diffraction grating theory. We don't go that far here, but we do attempt to impart a basic and functional knowledge of prisms and gratings.

Prisms

The prism and diffraction grating have long been used as dispersive elements within spectrographic instruments. What exactly is dispersion? It relates to the splitting of light, or electromagnetic radiation, into its constituent wavelengths, so clearly it is of great importance to the optical engineer; hence before we discuss prisms and gratings in detail let us look more closely at dispersion in general.

Dispersion

Dispersion is the variation of refractive index with wavelength exhibited by most optical materials. Typically, the refractive index of glass will fall with increase in wavelength. As an example, Schott BK7 has indices of 1.5302 and 1.5128 at wavelengths of 404.7 nm (n_h, violet mercury line) and 706.5 nm (n_r, red helium line) respectively.

Most glass companies will quote refractive index at a number of standard wavelengths. These are given various suffixes as defined in Table 3.1:

Table 3.1 Refractive index designation

Designation	*Wavelength nm*	*Spectral Line*
n_r	706.5188	red helium
n_C	656.2725	red hydrogen
$n_{C'}$	643.8469	red cadmium
n_d	587.5618	yellow helium
n_e	546.0740	green mercury
n_F	486.1327	blue hydrogen
$n_{F'}$	479.9914	blue cadmium
n_g	435.8343	blue mercury
n_h	404.6561	violet mercury

The form of the dispersion graph is usually well behaved over the visible region, but some glasses do exhibit a form of discontinuity in the refractive index function, and these glasses are said to exhibit anomalous dispersion; see Figure 3.1(a) and (b).

Dispersion coefficients enable a direct comparison between the dispersive performance of different glass types. The most common coefficient encountered is the Abbe V-number, also known as the reciprocal relative dispersion. The V-number is defined as follows:

$$V_d = \frac{n_d - 1}{n_F - n_C} \tag{3.1}$$

Taking Schott BK7 as an example again, we can either calculate the V_d value or read it from the catalogue entry; in either case it is given by

$$\begin{aligned} V_d &= (1.5168 - 1)/(0.008) \\ &= 64.17 \end{aligned}$$

Dispersion is the rate at which refractive index varies with wavelength, or, otherwise, the slope of the refractive index versus wavelength graph.

Generally glass types fall into two categories, flints and crowns. The flints exhibit high refractive index and high dispersion, while the opposite is true of the crowns.

A combination of glasses with different dispersion characteristics can serve to correct colour aberration within lenses intended for broad band applications. Thus a lens might be corrected so that the foci of two wavelengths coincide (achromatic correction). A more complex task is correction at three wavelengths with reduction of secondary spectrum; for this a number of different glasses with different dispersions are required.

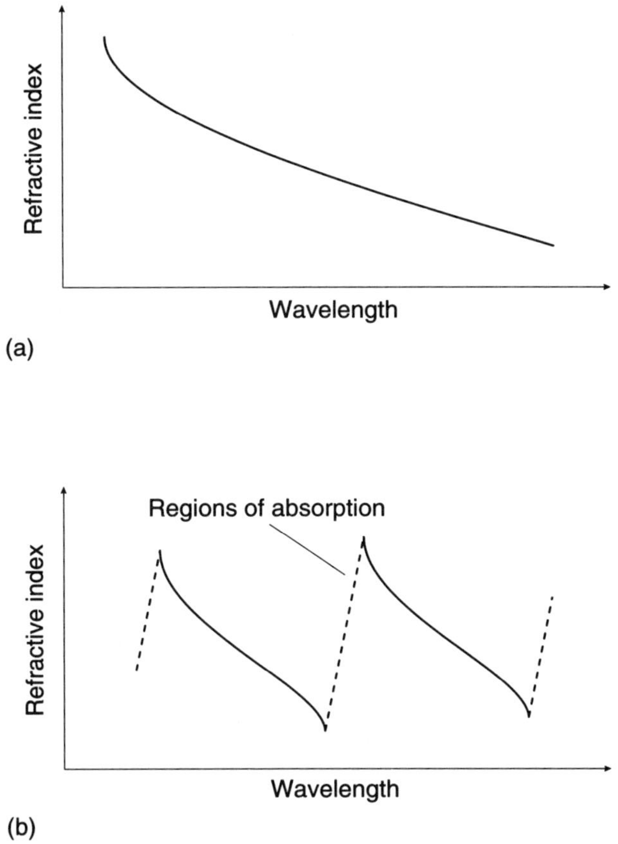

Figure 3.1 (a) Dispersion. (b) Anomolous dispersion

Types of prism

All prisms exhibit a degree of dispersion when used as refracting elements. When used in the configuration shown at Figure 3.2 the prism can be used, say, as the dispersive element of a monochromator or simply to produce a spectrum.

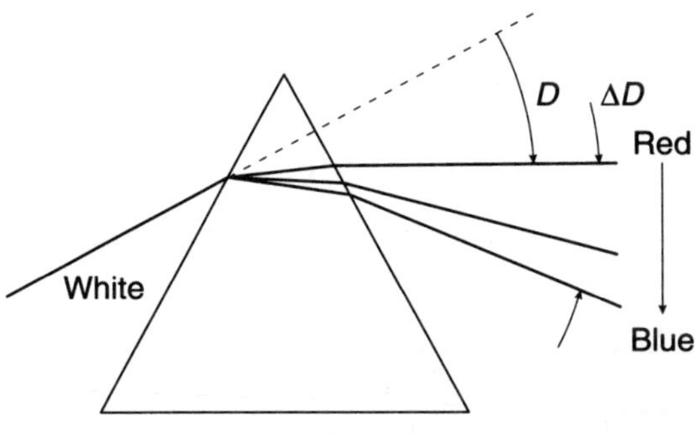

Figure 3.2 The dispersing prism (D=deviation of the ray through the prism)

The phenomenon of total internal reflection (TIR) is used within prism design and application. Total internal reflection occurs when a ray inside a block of glass, or other optical material, strikes the glass–air boundary at such an angle that the emerging, refracted, ray travels along the surface of the interface. Any increase in the incidence angle from that point results in the ray being reflected from the interface. No light escapes from this interaction and hence the ray is totally reflected.

Imagine a ray traversing the inside of a prism block and striking a glass to air boundary. The ray does not emerge from the block, but actually runs along the glass–air interface; that is, the angle of refraction is 90°. From Snell's law we can write

$$n_g \sin I = n_{air} \sin 90$$

$$\text{Sin } I = 1/n_g$$

If we assume the refractive index of air to be 1.00 and that the light to be of wavelength 587.6 nm (yellow helium line), and that the glass is BK7 with $n_d = 1.5168$, then

$$I = \sin^{-1}(1/1.5168)$$

$$= 41.25°$$

Thus, beyond an incidence angle of 41.25° the ray traversing the inside of our prism block will undergo TIR at the glass to air interface. This can be used to very great effect when steering light through 90° with a 45° prism.

It is interesting to note that TIR can be frustrated by placing a second air–glass boundary very close to the first or in contact with it. This is because a wave exists beyond the boundary as a consequence of the electromagnetic boundary conditions. This evanescent wave has an appreciable amplitude even in the case of perfect, 100 per cent TIR, but the amplitude falls to $1/e$ of its original value in a distance of about half a wavelength.

The sections that follow introduce some of the more commonly encountered prism types.

Right-angle

The right-angle prism is commonly used to deviate rays through 90° using total internal reflection (TIR) at the hypotenuse face. This means that, say, for a collimated ray bundle the entrance and exit faces of the prism are normal to the direction of propagation and dispersion is zero.

Right-angle prisms used in the above manner deviate the ray bundle through 90°, but introduce an image inversion (see Figure 3.3(a)).

When the hypotenuse face is used for beam steering through TIR it is usual to protect the face with a layer of paint or some such material. It is of course possible to aluminize the exterior face of the prism and use the face as an external reflector, to enhance the range of TIR below 42° or so.

Rhomboid

The rhomboid prism uses TIR much as the right-angle prism, but uses it twice to displace laterally the optical axis (see Figure 3.3(b)). The double reflection creates a single-axis double inversion, and therefore an erect and correctly handed image is transmitted.

Amici (roof)

The Amici or roof prism is in some respects similar to the right angled prism in that the input beam is deflected through 90 ° and an image inversion occurs. However, with the Amici prism the hypotenuse face is shaped like a roof (see Figure 3.3(c)). This has the effect of introducing a lateral inversion to the transmitted image as well as image inversion. In this way left goes to right and top goes to bottom, and a complete transform is achieved, unlike with the right-angle prism. The roof angle is such that TIR occurs for the lateral transfer.

The prism is often truncated to save on both weight and material, and various modified forms exist to enable an in-line function rather than round the corner.

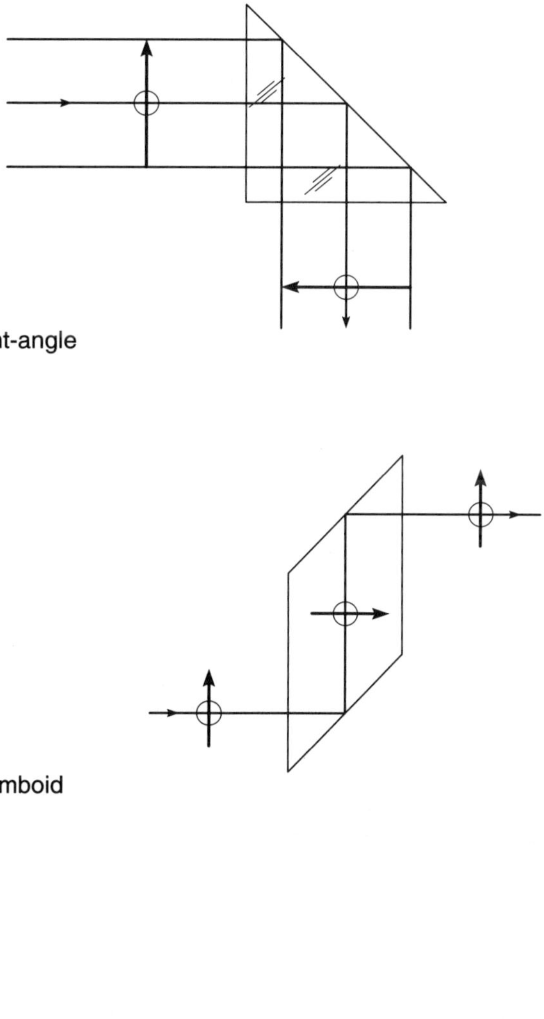

(a) Right-angle

(b) Rhomboid

Figure 3.3 Prisms

Continued

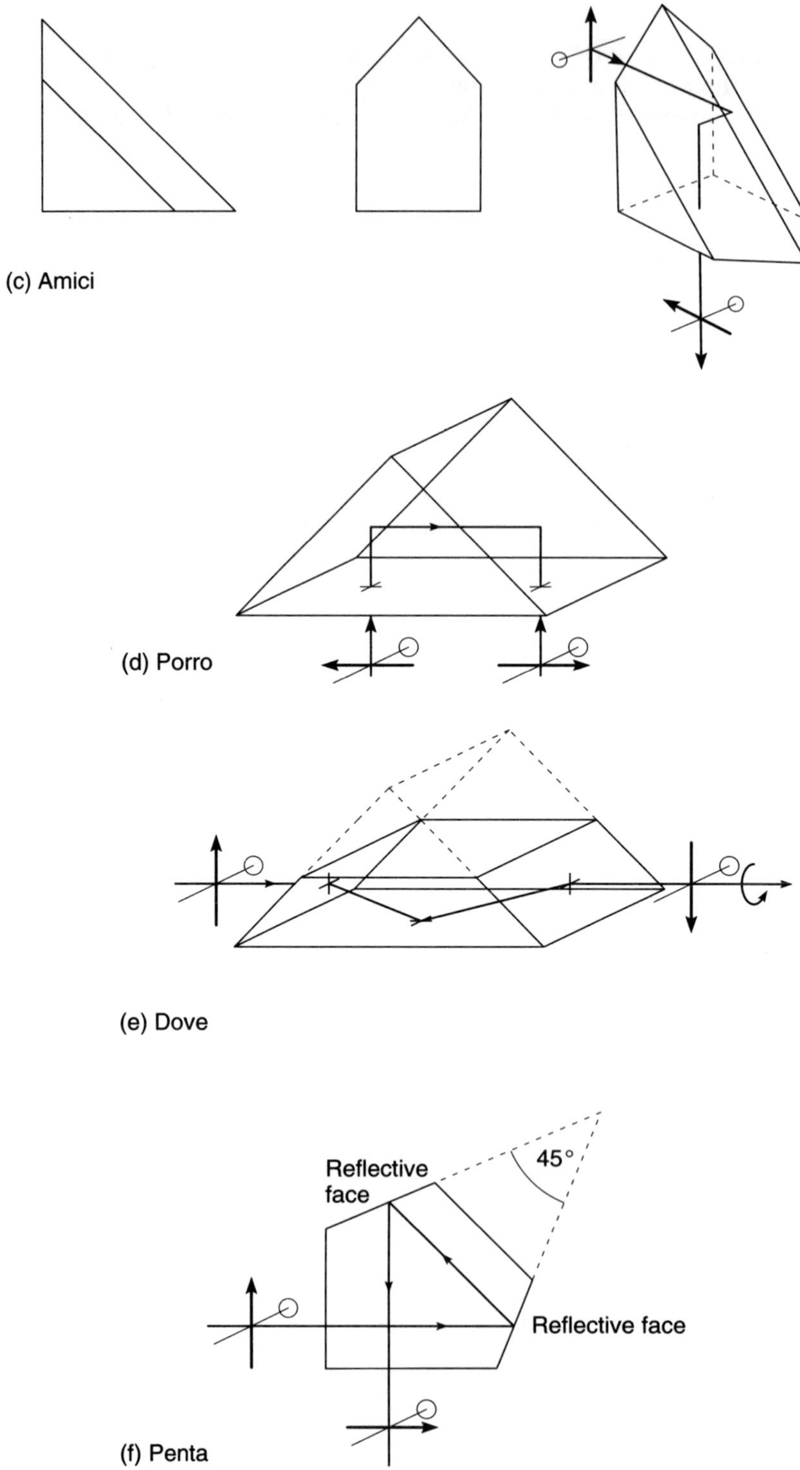

Figure 3.3 Prisms

Porro

A Porro prism can be considered as two right-angle prisms attached so as to form one large right-angle prism. The prism is used such that light enters at one side of the long hypotenuse face normal to the surface (thus minimizing dispersion effects). This light is totally internally reflected by the two short faces and emerges from the opposite end of the hypotenuse face (see Figure 3.3(d)). The beam is steered through 180 °.

The image undergoes an inversion while traversing the prism. Traversing two such prisms erects the image, a technique used commonly in binocular design.

Dove

The Dove prism is a truncated right-angle prism. The optical axis of the system is arranged such that it is parallel with the hypotenuse of the prism but displaced such that the beam strikes one of the short faces. The beam refracts into the prism, reflects via TIR at the hypotenuse and then emerges from the second short face (see Figure 3.3(e)).

The Dove prism exhibits the interesting characteristic of rotating its image at twice the rate at which the prism can be rotated. The prism returns its best performance when operated with collimated light.

Penta

The Penta prism turns an image through 90 ° without altering the image orientation. This is particularly useful for photographic camera systems and for the end steering blocks on optical rangefinders. These prisms have also been used as optical squares for surveying or in machinery alignment applications.

Figure 3.3(f) shows the layout of the penta prism with the two reflecting surfaces silvered because some landing angles will not exceed the critical angle for TIR. The angle between the two reflecting surfaces is 45 °.

Cube beamsplitter

There are several ways in which an optical beam can be split from one to two channels. These 'beamsplitters' come in several different guises, which include plate, pellicle (thin membrane) and cube. The cube beamsplitter is in essence a prism and thus deserves a mention here.

A cube beamsplitter (Figure 3.4), is manufactured from two right-angle prisms

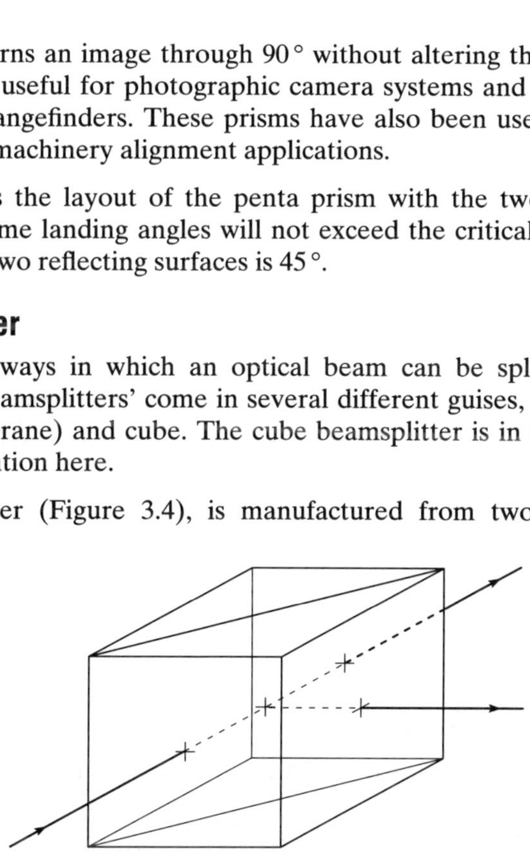

Figure 3.4 Cube beamsplitter

cemented together along their hypotenuse faces. One of the hypotenuse faces has deposited on it a partially reflecting, partially transmitting coating that operates efficiently at incidence angles of 45 °.

With care in the manufacture and use of a high-performance coating, beamsplitting ratios of very nearly 50:50 can be achieved. However, the beamsplitter coating can be manufactured to provide virtually any ratio from 5:95 through to 95:5. No system is perfect and some losses do occur, particularly if the entrance and exit faces are not antireflection-coated, in that instance splitting ratios of 40:40 can be envisaged.

Wollaston prism (polarizing prisms)

The Wollaston prism serves to separate the two orthogonal polarization states of light by using a prism made of a birefringent material such as calcite ($CaCO_3$).

Light traversing a birefringent material will experience a different refractive index according to its state of polarization. The ordinary or 'o' rays will experience a given refractive index and behave according to the laws of refraction and reflection. However, the extraordinary or 'e' rays will, according to direction of travel and wavelength, experience a completely different value of refractive index. The o and e rays are effectively orthogonal linear polarization states.

The difference in refractive index seen by the o and e rays can be large. Calcite, a commonly used birefringent optical material, has a difference in index of about 0.15 with the o ray having the higher index and hence slower propagation velocity. The crystal retains a dispersion characteristic, with refractive index magnitude varying with wavelength for both the o and e rays. Hence polarizing prisms work best with monochromatic light.

Calcite has the o ray experiencing a higher refractive index than the e ray, whereas crystalline quartz exhibits a higher index for the e ray. Where the extraordinary refractive index is higher than the ordinary the material is known as optically negative according to ($n_o - n_e$=negative).

The extraordinary refractive index varies with propagation angle and coincides with the ordinary refractive index along the optical axis of the material.

If randomly polarized light is incident at an angle to a birefringent material boundary then the o and e rays will experience a different magnitude of refraction, thus separating the o and e rays. This is the principle of the Rochon and Wollaston prism types.

Effective beam separations from about 15 ° up to a maximum of about 45 ° are commercially achievable with the Wollaston type of polarizing prism. If the two emergent beams overlap a little then that region of overlap will remain randomly polarized.

The angle at which total internal reflection occurs is also dependent on refractive index and can therefore be used to divide the o and e rays, as with the Nicol and Glan–Thompson types of prism.

Figure 3.5(a) and (b) illustrate the operation of the Wollaston and Nicol polarizing prisms.

For practical purposes the prism type polarizers are efficient, but usually heavy and relatively expensive. Sheet polarizers are light and inexpensive, but do not perform so

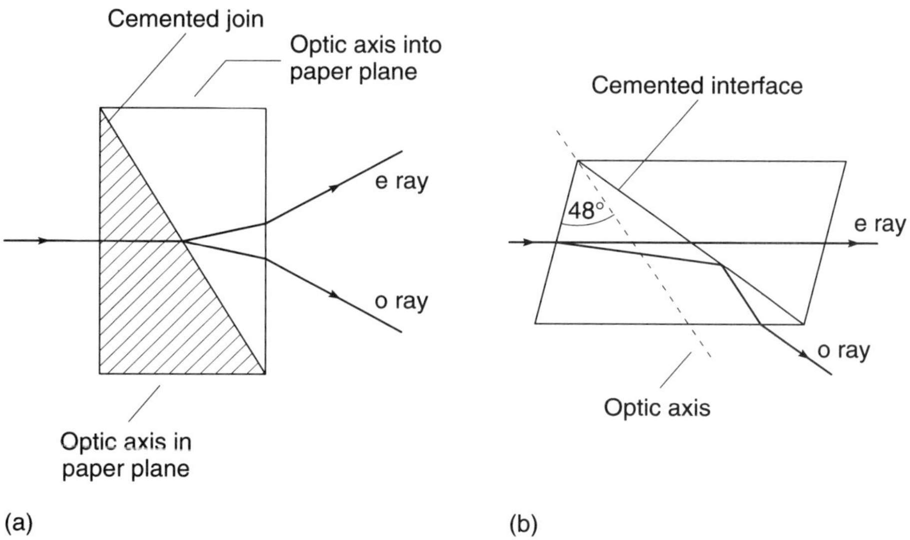

Figure 3.5 *(a) Wollaston prism, (b) Nicol prism*

well as the prism types. The prism types usually perform over a larger wavelength range than do the sheet types. The useful wavelength range for a Wollaston prism made from calcite is of the order 350 to 1500 nm, whereas sheet polarizers are usually restricted to the visible or near infrared.

Polarizing beamsplitters

Polarizing beamsplitters, which are manufactured from two cemented right-angle prisms with a dielectric coating between, operate such that light is incident at the Brewster angle. This arrangement means that the p-polarization is transmitted with virtually no reflection and the s-polarization is reflected.

The very thin nature of the dielectric coating – a few micrometers only – results in a virtually 'ghost' free reflection.

The dielectric coating mimics a stack of plate polarizers, each at the Brewster angle, all transmitting the p-polarization and reflecting the s-polarization. A larger number of plates, or stacks, improves the isolation between the s and p states.

Normal beamsplitters lose 3 dB on recombination of two beams. However, the application of a polarizing beamsplitter avoids this optical loss. A particular application is in interferometry, where the reference and sensing beams can be recombined without significant loss provided that they are orthogonally polarized.

Types of diffraction grating

The diffraction grating has found many uses, spanning a long and serviceable history of development from the very first ruled examples through to complex holographic elements.

In this section the basics of diffraction gratings are introduced, the two types, reflecting and refracting, are discussed and finally a few words are devoted to their manufacture.

The diffraction grating is of course another dispersive element, causing light of different wavelengths to behave differently. The diffraction grating really is the engine of most spectrographic instruments, and much effort has been directed at producing gratings of ever-improving quality. The production of good-quality gratings is now very reliable, with modern replication techniques such as hot-foil embossing of holographically produced masters.

Once again we start the section with a review of the basics.

Basics

Diffraction gratings are essentially dispersive elements. A basic grating will divide the light power between a number of different diffraction orders, each providing a spectrum of the incident light. Some diffraction gratings can preferentially put most of the optical power in one order only and are referred to as blazed gratings.

Diffraction by a regular array is described by the following simple formula:

$$d \sin \theta_n = n\lambda \tag{3.2}$$

where d = array spacing
θ_n = angle through which ray is diffracted for normal incidence
n = order of diffraction
λ = wavelength of incident light.

For non-normal incidence of the ray the following form must be applied:

$$d \sin(\theta_n - \theta_i) = n\lambda \tag{3.3}$$

where θ_i is the incidence angle of the light (see Figure 3.6).

As an example, consider a monochromatic light source of wavelength 500 nm incident normally on a transmission grating with 1250 lines per millimetre. What is the angle of the first diffracted order?

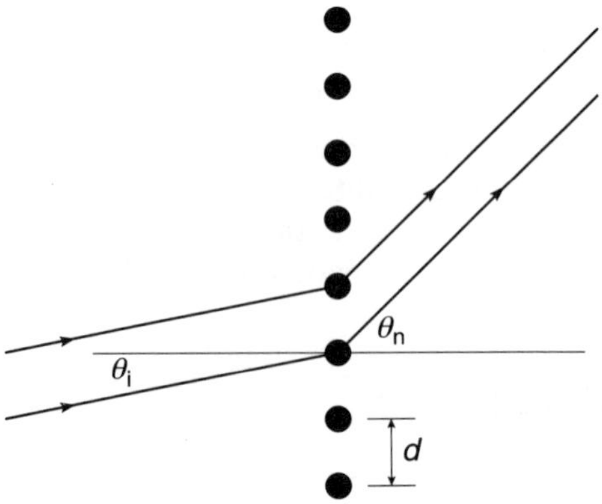

Figure 3.6 Diffraction grating. $d \sin(\theta_n - \theta_i) = n\lambda$

The line spacing will be 1/1250 = 8×10^{-7} metres. From above, at normal incidence $\theta_i = 0$. Rearranging gives

$$\theta_n = \sin^{-1}(n\lambda/d) \tag{3.4}$$

We have $\lambda = 500\,\text{nm}$, $d = 0.8\,\mu\text{m}$ and we are interested in the first order so that $n = 1$. Thus

$$\begin{aligned}\theta_1 &= \sin^{-1}(1 \times 500 \times 10^{-9}/8 \times 10^{-7})\\ &= 38.7°\end{aligned}$$

If the diffracting angle exceeds 90° for normal incidence then the diffracted order cannot exist. This is called an over-the-horizon diffraction order.

Consider now the same grating with white light striking it at normal incidence. Take each end of the white-light spectrum, say, 400 nm and 700 nm, and calculate the first-order diffraction angles, as follows:

400 nm:
$$\begin{aligned}\theta_1 &= \sin^{-1}(1 \times 400 \times 10^{-9}/8 \times 10^{-7})\\ &= 30°\end{aligned}$$

700 nm:
$$\begin{aligned}\theta_1 &= \sin^{-1}(1 \times 700 \times 10^{-9}/8 \times 10^{-7})\\ &= 61°\end{aligned}$$

From this analysis it seen that the longer wavelengths are steered through a larger angle than the shorter ones.

The resolving power of a grating, which quantifies the ability of a grating to resolve two adjacent wavelengths, is given by the following, which is based on the Rayleigh principle:

$$\lambda/\delta\lambda = nN \tag{3.5}$$

where λ = wavelength
$\delta\lambda$ = minimum separation of next wavelength that is just resolved
n = diffraction order
N = total number of grooves.

The larger the number of grooves, the better the resolving power of the grating.

This analysis assumes that the grating is the limiting aperture of the system. The total number of grooves within a diffraction grating is given by

$$N = W/d \tag{3.6}$$

where W = the full width of the grating and
d = the grating pitch (distance between grooves).

Thus the resolving power of the grating depends on the total width or size of the grating. Therefore accurate spectrographic work requires large diffraction gratings.

The resolving power is a dimensionless quantity, and to indicate order of magnitude the following example is offered: wavelength = 500 nm, $n = 1$, $W = 25\,\text{mm}$, $d = 1\,\mu\text{m}$ (1000 lines per mm).

The resolving power is given by nN, which in this case is given by

$$1 \times 25 \times 1000 = 25000$$

What minimum separation of wavelength can just be resolved?

From the above,

$$\begin{aligned}
\delta\lambda &= \lambda/nN \\
&= 500 \times 10^{-9}/(1 \times 2.5 \times 10^{4}) \\
&= 2 \times 10^{-11} \\
N &= W/d \\
&= 25 \times 10^{-3}/1 \times 10^{-6} \\
&= 2.5 \times 10^{4}
\end{aligned}$$

Wavelengths separated by about 0.020 nm will be just resolved. This situation can be improved byany of the following:

- Doubling the width of the grating $\delta\lambda = 0.010$ nm.
- Selecting fourth diffracted order $\delta\lambda = 0.005$ nm.
- 10 000 lines per mm $\delta\lambda = 0.002$ nm.
- All three $\delta\lambda = 0.00025$ nm.

Thus good resolving power is achieved. However, it is important to differentiate between grating resolving power and the resolution capability of the final instrument. This will depend on a number of factors, such as slit width for a monochromator, and will need to be determined separately.

However, the creation of a 50-mm square grating with 10 000 lines per millimetre is not really practical.

The diffraction grating, transmitting or reflecting, is commonly used in collimated light just as a dispersion prism would be. Thus, a collimating lens is required with a small source to produce a well-collimated beam. The diffracted light is then usually collected by an objective that brings the infinitely distant image to a focus.

Transmitting

This type of diffraction grating works in transmission, and literally comprises a large number of opaque and clear stripes. Some of the light incident on the grating is unaffected by the stripes, and some is diffracted. Therefore a large amount of optical power is usually observed in the zero, non-diffracted, order.

Transmission-type gratings are usually manufactured on a plane glass or plastic substrate. This type of grating is therefore generally limited to the visible and near-visible region of the spectrum, owing to the absorption characteristic of the substrate. Gratings for use outside the near-visible region are usually reflective when the selection of the coating and the material can give a good efficiency of reflection.

Applications for transmission gratings include the conversion of telescopes or cameras for employment in spectrography astronomy. Material considerations typically limit application to the near-visible wavelength range of between about 0.2 μm and 1.2 μm. Line frequencies do not commonly exceed 1000/mm.

Reflecting

This type of grating works in reflection and comprises a mirror-like substrate of grooves or facets. The spaces between the grooves are non-reflecting.

To simplify some instruments the reflection grating can be formed on a spherical or other shape of substrate, so that the component itself has some optical power. For example, this approach could be used with a monochromator such that the grating forms an image of the input slit at the detector.

Blazed reflecting diffraction gratings are those that have the reflecting facets angled such that both the specular reflection and the first diffracted order lie in the same direction. In this way more optical power is obtained in the first diffracted order.

The amplitude grating

The amplitude, or line, grating is simply an alternative way of describing both of the grating types discussed above. The wavefront, transmitted or reflected, undergoes modification in amplitude due to the grating.

The grating lines can have sharp or smooth boundaries. The classic transmission grating has sharp boundaries, whereas a ruled reflection grating will form grooves according to the tool shape and depth of cut.

Gratings with an amplitude transmittance that follows a cosine-squared function are referred to as sinusoidal gratings and produce only two spectra, positive/negative first-order and the zero-order. Hence, with careful selection of the ratio of groove depth to spacing, sinusoidal gratings can provide useful amounts of power in the first diffracted order.

Amplitude grating efficiency is usually quite low. With the transmission type, 50 per cent of the grating area is opaque and hence half of the incident power is lost. If we add to that loss bulk absorption, scatter from imperfections and perhaps some dust and grease, then efficiencies can fall to quite low levels.

Many high-performance and relatively low-cost amplitude gratings are made through the replication of a holographically produced master. This method is discussed below under 'Manufacture'.

The phase grating

Traditional reflective and transmissive gratings produce ordered spectra according to the principle of phase addition. Waves in phase add amplitudes and waves of equal amplitude in antiphase cancel.

Phase changes can be introduced across a wavefront by arranging for it to traverse a region of varying optical thickness. Thus a regular periodic structure that introduces a phase shift to a wavefront will act as a diffraction grating. Figure 3.7 represents a two-

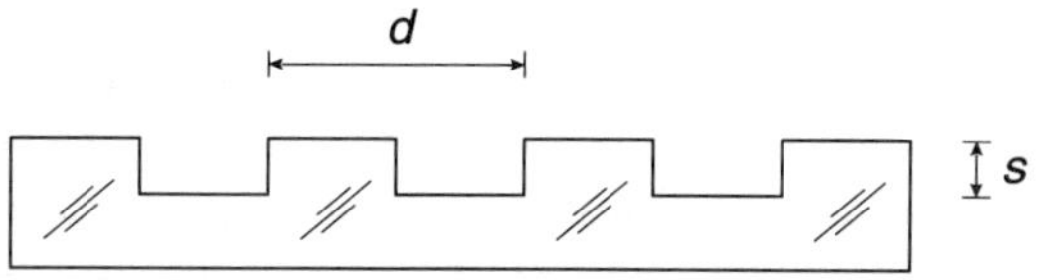

Figure 3.7 Phase grating. Optical thickness of grooves = nS. When nS is equal to λ/2, the waves cancel. When nS equals 0, only the zero order is transmitted; this is a window.

dimensional structure of optically transparent material. The light that traverses a 'thin' region will undergo a different phase shift from that undergone by light that traverses a 'thick' region. The relative phase shift is controlled by the optical thickness nS. For example, when the product nS equals half a wavelength the waves will be 180° out of phase and will cancel. When the product nS is zero there will be no diffraction and the grating will behave as a window.

There are no opaque strips with a phase grating and hence optical efficiencies can be kept high, at best to the order of 90 per cent.

If a phase grating of circular symmetry is produced then a focusing effect is observed. This is the phase or diffractive lens. Such a lens will exhibit several foci corresponding to different orders of diffraction.

This section has so far described the transmission phase grating. A similar principle can be applied in reflection, where the step height of the now reflective facets are of the order of a fraction of a wavelength. This is the reflective phase grating, manufacturing techniques for which could include photoresist and etching, or, for longer wavelengths, diamond machining.

Phase gratings and lenses as briefly described above are sometimes referred to as binary optics because of the stairlike pattern that approximates to the desired profile. The interest in these elements stems from the reliability of now mature manufacturing technologies such as laser writing and diamond machining. The resulting elements can have weight and cost advantages. Some of the more exciting applications include correction of vision defects through binary contact lenses and photonic (electro-optical) circuits such as might be used for optical computing.

Stray light

Any imperfections contained within a diffraction grating will direct light in other than the designed or predicted directions. This can in some instances produce lower than expected signal-to-noise levels at the system detector. This is important in such systems as, for example, a simple monochromator operating in the ultraviolet region and having both the source and the detector operating at the limit of their performance. The presence of stray light at these wavelengths will severely limit the reliable detection capability of the instrument. Note that for this particular example stray light will also occur from the mirrors and walls of the monochromator.

A holographic diffraction grating produces less stray light than a ruled diffraction grating. This is due to a very regular spacing of grooves that are smooth and uniformly shaped.

Manufacture

Traditionally, diffraction gratings were manufactured individually through the application of a ruling engine which used a diamond stylus to scribe the lines. The mechanical demands on such machines are very high, with close tolerances required on straightness of line and general finesse. Interferometric control of the ruling arms, together with thermal and vibrational control of the environment, are required to achieve the very best in grating quality. This approach is not practical for the mass production of general purpose gratings, nor for the lower volumes of research or instrumentation gratings. Hence accurate replication methods have become acceptable for most grating applications.

Various techniques have been explored, which include:

- A master grating, produced via a ruling engine, coated with a hard material and converted to a plastic moulding tool from which many replicated gratings can be produced.
- Highly accurate gratings, produced through holographic techniques, can be exposed directly on a photoresist material. The photoresist is exposed, etched and nickel-plated to produce a replication tool. This is a relatively fast process compared with the several days required to produce a single ruled grating, and the gratings replicated from the tool are at a much lower cost.
- Direct laser writing or electron beam lithography with line widths of between 100 nm and 5 μm can be used to generate gratings. Nickel shims are then produced through electroplating and the resulting tool used for hot foil embossing.

Table 3.2 gives an indication of the structure size that can be generated with three different generation techniques. Shim fabrication is achieved by electroplating nickel onto an original surface relief structure produced through one of the techniques described above. The original surface is first coated with the deposition of a thin silver, gold or nickel film. A nickel foil of up to 100 μm thick is then electroplated and separated from the structure to achieve a first-generation replication shim. Second-generation foils can be manufactured from the first in a similar way, and it is usually this shim that will be used in the replication tooling. The first generation shim will be kept as a master to enable further tool manufacture as and when required.

Table 3.2 Grating structure sizes available

Process	*Structure size μm*
Laser interference	0.2 to 2
Electron beam writing	0.1 to 5
Laser beam writing	0.5 to 5

Replication is achieved in polymers by the direct hot embossing or moulding of foils. The direct embossing of polymer films on glass substrates has been demonstrated, and holds the promise of very stable and highly accurate gratings at low cost.

These manufacturing techniques are not limited to the replication of diffraction gratings but can be applied to decorative foils, holograms, integrated optic and other waveguide structures.

Diffractive optics

We are used to the concept of refractive and reflective optics (lenses and mirrors), and we are now familiar with dispersive systems such as prisms and diffraction gratings, but what are diffractive optics?

The Fresnel zone plate is a form of diffractive lens. This is a plane disc that comprises alternate opaque and clear zones and forms a focus on axis when illuminated by a wavefront. Interestingly, diffraction theory gives the surprising result that in the centre of the shadow of an opaque circular disc there will be a bright spot. The following analysis gives some insight as to why this is so.

Diffractive lenses are associated with similar principles to the zone plate, so it is worth a few paragraphs explaining just how a zone plate works. Consider Figure 3.8. Here A is a monochromatic point source of spherical wavefronts (recall that a wavefront is a surface of constant phase) and B some point on the axis beyond the zone plate. The plate is divided into zones bounded by circles of radii OP_1, OP_2, etc.

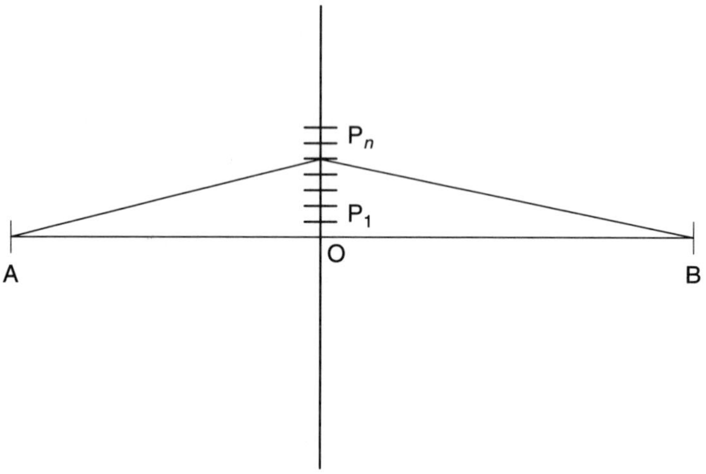

Figure 3.8 Zone plate

Let the geometry of this configuration be such that the optical paths are defined by

$$AP_1 + BP_1 = AO + BO + \lambda/2$$
$$AP_2 + BP_2 = AO + BO + 2\lambda/2$$

or, for *n* zones,

$$AP_n + BP_n = AO + BO + n\lambda/2 \qquad (3.7)$$

These are referred to as half-period zones. As the wavefront traverses each zone the phase at B is displaced by 180° ($\lambda/2$), causing alternating destructive and constructive interference at B. Now, make each alternate zone opaque and the light arriving at B is in phase and constructively interferes. This is a lens!

The lens is not very efficient and only works well in monochromatic light, but it *is* a lens. There is no light bending due to refraction involved; this bending is due to diffraction at the circular aperture (recall that the performance of a perfect lens is limited by diffraction at a circular aperture).

Take Equation (3.7) and use the Pythagoras theorem to put the distances AP_n and BP_n in terms of r_n, the radius of the nth zone, apply a binomial expansion and the following expression is achieved:

$$r_n^2(1/AO + 1/BO) = n\lambda \tag{3.8}$$

Rearrange this equation and we find that r is proportional to $n^{1/2}$. The radius of each zone is proportional to the square root of a natural number, 1, 2, 3, etc. One stage on, with a little more rearranging, we find

$$1/AO + 1/BO = n\lambda/r_n^2 \tag{3.9}$$

This is identical to the thin-lens formula for a lens with focal length of $n\lambda/r_n^2$. So, given fixed conjugate conditions you can determine the zone radii to produce a lens of the desired focal length. This focal length is inversely proportional to the wavelength, so a single zone plate will only form an image over a very limited spectral bandwidth. However, many applications are laser-orientated, and this limitation does not then pose a problem.

A few final words with respect to the zone plate should include the fact that it has more than one focal length. For example, waves from the third, seventh and eleventh zones add together to form a focus three times larger than when all of the zones add together.

Before we go much further with this discussion, we should point out very clearly that the Fresnel zone plate or diffractive lens is an optical element that is very different from the Fresnel lens. This is indeed confusing, but that's the way it is! The diffractive lens interferes coherent light to produce a focus, while the Fresnel contributes incoherently to produce an image.

To improve the efficiency of the zone plate, or diffractive lens, instead of blocking alternate zones these can be made of such a thickness that the wave is retarded by half a wavelength. This is of course similar to the way in which the phase grating works.

Taking this concept a stage further, the diffractive lens can be made more efficient by encoding phase information in the physical profile of the zone. Such a structure is called a kinoform. Again let's be clear: a kinoform achieves phase modulation of a wavefront through surface relief profile; this has nothing to do with refraction! The kinoform profile performs the same function as the blaze angle on a diffraction grating, and the relief profile is only about two wavelengths in height.

Look at Figure 3.9. A zone plate is illustrated as a series of concentric zones alternating light and dark. You can see at once that this is not a very efficient component in terms of clear aperture alone. Three forms of kinoform are illustrated. The quadratic profile is the most efficient, with a theoretical maximum of 100 per cent compared with the amplitude zone plate at about 5 per cent. Next comes the linear profile at about 99 per cent and finally the binary profile at about 81 per cent for a four-step profile; the inclusion of more steps increases the efficiency.

In practice, of course, a 100 per cent efficiency will not be achieved. A detailed diffraction analysis will reveal some loss mechanisms according to the physical geometry of the zones. There will be Fresnel reflection losses for uncoated elements together with bulk absorption losses.

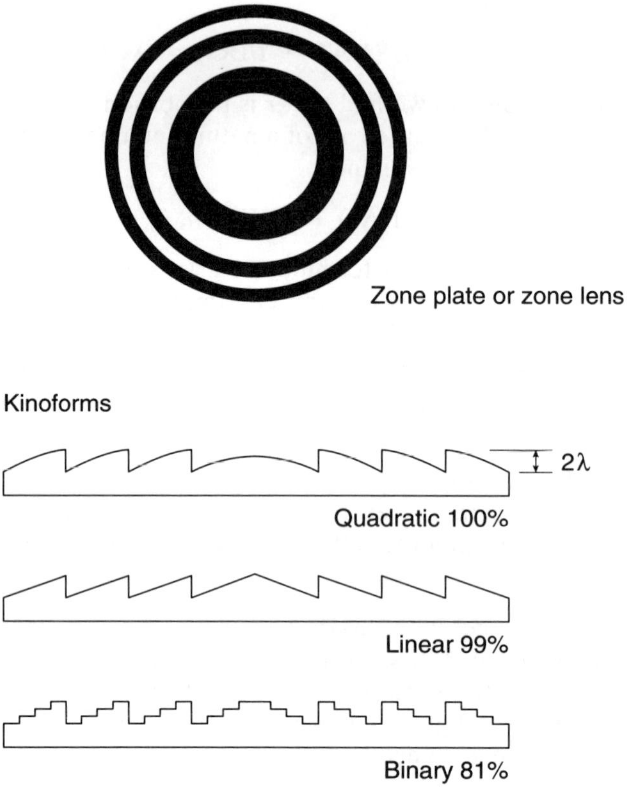

Figure 3.9 Diffractive lens

The profiles presented in Figure 3.9 can be fabricated through diamond turning techniques or photofabrication with masking, exposure and etching operations. Holographic manufacturing techniques can be applied with very great accuracy, which leads to the interesting phenomena of diffraction-limited diffractive lenses. Replication techniques can be applied that are identical to those discussed above in the section on types of diffraction grating, giving the potential for low-cost elements of light weight and good accuracy.

Where will diffractive optical elements find application? These elements are ideally suited for laser applications, where light weight, low cost and high quality are important. The head-up display (HUD) is one obvious example where quality and weight are important.

CHAPTER 4

Reflectors and mirrors

Introduction

This chapter aims to introduce the reader to some of the issues that govern the selection of mirrors or reflectors for specific tasks. Aspects covered include those of surface quality and specification, manufacturing techniques and some of the rules that govern application.

The application will enable specification of the particular mirror type, substrate material and coating, and some guidelines are presented to enable a rapid identification of the right solution.

Anyone who is getting ready to specify a mirror for the first time should ask the following questions:

- How good a surface finish do I need?
- What thermal stability or capability do I require?
- What reflectivity do I need?
- How much can I afford to spend?
- How robust a mirror do I need?
- How do I want to mount the mirror?
- Will the coating be suitable for the spectral region of my interest?

Answers to these questions and more will be found in this chapter.

Optically flat mirrors

The term optically flat mirror, or optical flat, relates to a surface where the departure from a plane is small compared with the wavelength at which the flat is to be used. This departure, for true optical surfaces, is a measure of surface form and not roughness.

For most laboratory or simple beam-steering tasks a flatness of, say, ten wavelengths (total form error not to exceed ten wavelengths) is adequate. However, for a demanding task such as steering within an image-forming system, a flatness of one-tenth of a wavelength may be required.

For the most demanding interferometric applications the very highest quality mirrors may be required with flatness of form approaching one-hundredth of a wavelength.

Flat mirrors are usually described in what is becoming a standard form (Table 4.1).

Table 4.1 Mirror definitions

Description	*Flatness*	*Surface finish*
Demanding laser systems	$\lambda/20$	10/05
Interferometric quality	$\lambda/20$	40/20
Research quality	$\lambda/10$	60/40
Laboratory quality	$\lambda/4$	60/40
Utility	λ to 5λ	80/50

The surface finish is usually described by two numbers, referred to as scratch and dig; these are defined below.

- *First number, scratch.* Indicates the apparent width of a scratch. Specified against standard scratches, such as a #80 standard scratch.
- *Second number, dig.* Indicates the apparent diameter of a dig, bubble, inclusion or whatever. A number, say 50, indicates a defect diameter of 0.5mm, 20 a diameter of 0.2mm and so on. Irregular digs are sometimes quantified by the sum of half the length plus the width.

These two numbers form a surface defect code that is readily interpreted by all optical workers. The code is usually accompanied by an indication of the allowed number of defects on or in the component. Inspection of components for scratch and dig is accomplished visually against known reference standards. Thus an 80–50 surface would be fairly easy to fabricate and therefore inexpensive. A 20–10 or 10–5 surface is much more difficult to fabricate and is limited to demanding applications only, such as an illuminated reticle. If a surface flat to $\lambda/50$ and having no more than five, 10–5 defects over the full surface area, is specified then you must be prepared to spend a significant amount of money!

It must be stressed that surface defect counting is a highly subjective activity, and is subject to operator and standards variation.

Surface roughness is not usually a problem with an optical flat, because optical surfaces are usually very smooth. This smoothness is a consequence of the manufacturing techniques applied. However, some processes, such as diamond turning, will leave an ultra-fine micro surface structure that needs to be quantified.

Surface roughness can be specified in several different ways. One common specification is called the centreline average method (R_a). It is the average height from a mean line of all heights of the surface, whatever the sign of the he.ght (see Figure 4.1):

$$R_a = \frac{h_1 + h_2 + h_3 + \cdots + h_n}{n}$$

$$= \Sigma\, h/n$$

Alternatively, some manufacturers may specify a peak to trough variation, whilst others may use a ± departure from the average. It is always advisable to check which convention is being used.

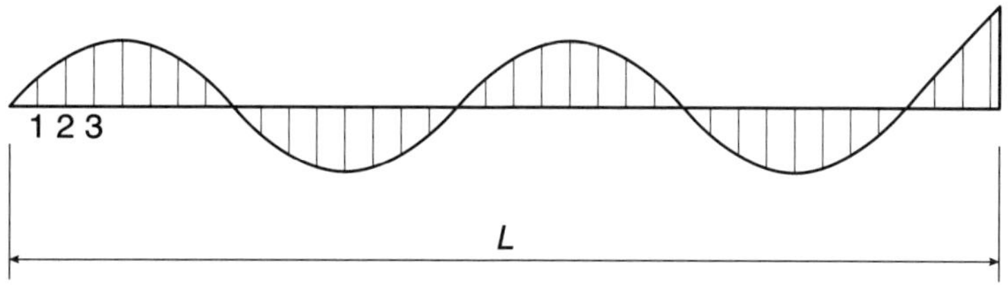

$$R_a = (\Sigma h) / n$$

Figure 4.1 Centreline average method (R_a)

With reference to Figure 4.1, it is usual to specify the area over which the flatness or roughness specifications are to apply. This is indicated by the dimension L that, for example, represents the clear aperture of a mirror.

There are many ways in which the laborious task of measuring such an ordinate set can be avoided. The most common is to use a machine.

Surface roughness is of great interest to the long-wavelength (infrared) optical worker. Here, at a wavelength of, say, 10 µm surface deformations of about a wavelength are 20 times larger than for visible wavelengths. Thus, a relatively crude surface can be used with no loss of performance. A surface with such a large microstructure roughness applied in the visible wavelength range will act as a diffuser.

Fabry–Pérot etalon

An etalon is a pair of plane, partially reflecting mirrors that have their reflecting sides facing each other so that light will reflect back and forth between the two. The gap between the two parallel plates need not be air, and in practice various materials are used. For example, a laser cavity is an etalon created by the cavity mirrors.

The Fabry–Pérot etalon (FPE) is really an optical resonant cavity for those wavelengths where the optical path between mirrors is an integer number of wavelengths. FPEs can be thought of as multiple-beam interferometers that support only those wavelengths that are an exact fraction of the optical path of the cavity.

If there is no phase difference between emerging wavefronts, after one, two or many reflections within the etalon, then constructive interference occurs and the transmitted wavefront is a maximum. This only occurs when the cavity optical path is an integer number of wavelengths. For all of the other wavelengths destructive interference minimizes transmission, with most of the light reflected back toward the source.

For constructive interference and a maximized transmission the following condition applies:

$$m\lambda = 2nd\cos\theta$$

where m = an integer referred to as the order
λ = wavelength of transmission maximum
n = refractive index of cavity
d = thickness of cavity
θ = angle of incidence to cavity.

When the etalon is placed in a beam of monochromatic light a series of circular fringes is produced. This is the result of multiple interference, owing to the partial transmission of the etalon at each reflection on the exit side.

The FPE can be used as a laser cavity tuning device where the laser resonance is restricted to that of the FPE. This is a very effective line tuning or bandwidth narrowing device, and is used in modern telecommunication lasers to reduce linewidth. Tuning or spacing of the FPE plates can be achieved by tilting the device and thus effectively increasing, or decreasing, the optical path. FPEs find application in spectroscopic applications where very narrow bandwidths are required. The principles behind FPEs are regularly applied in the field of narrow-band coatings.

With careful control of the spacing between etalon plates, a high finesse of wavelength discrimination can be achieved.

Curved mirrors

Curved mirrors, or mirrors with optical power, fall into several main categories. These are:

- Spherics
- Parabolas
- Ellipses
- Free form
- Concentrators

These categories of mirror find application in many diverse areas where characteristics such as the following are important to the success of the application.

- Achromatism
- Simplicity
- Light weight
- Broad wavelength range
- High-power handling capability

These characteristics may be required individually or collectively. For example, a high-power laser beam delivery system for industrial welding will require high-power handling capability to the extremes of cooling the mirror substrate. However, even a simple monochromator will require achromatism over a wide bandwidth.

The relative merits and areas of application for curved mirrors are discussed in the following sections. Not discussed are the specialist technologies of adaptive mirrors or aperture-synthesis applications, where individual mirrors are linked together to form very large structures. This is a very exciting area of mirror technology, but well outside the scope of this book.

Spherics

Owing to its ease of manufacture the spheric is historically one of the most common optically powered mirrors. It finds application across many fields, from imaging and condenser optics through to high-power laser systems. One major advantage of it is that with the correct selection of materials a high reflectivity can be maintained across very large bandwidths extending from the ultraviolet through the visible and far infrared.

One drawback of the spherical mirror is spherical aberration, which has an aperture dependence. This aperture dependence relates directly to speed (f-number), and therefore spherical mirrors used in imaging applications tend to be of relatively high f-number, i.e. slow. This aperture restriction also acts to limit light-gathering power.

The effective focal length of a spherical mirror is given by the simple expression

$$f = R/2$$

Where f is the focal length and R the radius of curvature of the surface. This relationship is illustrated in Figure 4.2 and only applies with any accuracy to the paraxial region.

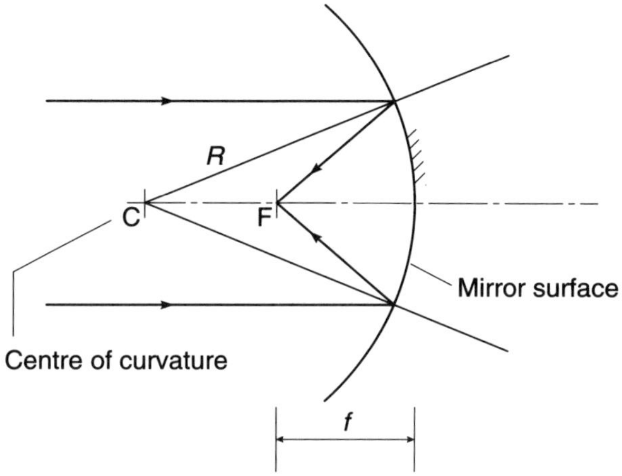

Figure 4.2 Spherical mirror focus

The mirror equation, relating object and image positions, for the paraxial region only is given below.

$$1/u + 1/v = 2/R = 1/f$$

where u and v are the object and image distances and are interchangeable (see Figure 4.3). Thus, for an object at a distance $u = 10$ cm from a spherical mirror of radius 100 cm the paraxial focus will be located at

$$(2/100 - 1/10)^{-1} = -12.5\,\text{cm}$$

Note that for a concave mirror a real image is formed on the same side of the mirror as the object. However for a convex mirror a virtual image is formed beyond the mirror surface.

The definition and representation of magnification for spherical mirrors follows very closely that of paraxial optics for spherical lenses. That is, the ratio of object to image

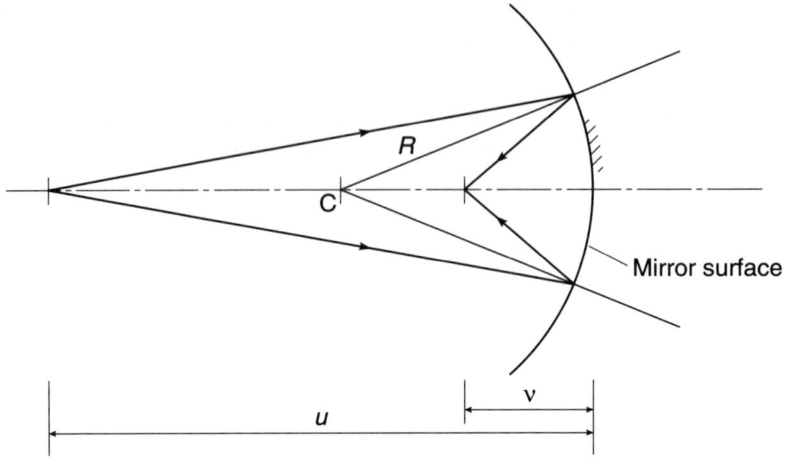

Figure 4.3 Spherical mirror focus

height or image to object distance will provide a measure of the magnification. However, again, these relationships are strictly limited to the paraxial region.

The fields of application for the spherical mirror are many and numerous. As an example of application we can mention the projection system condenser. A schematic of a classic projector is illustrated at Figure 4.4.

The key components of this system are the condenser mirror, the light source, the condensing lens, the film gate or object to be projected and the projection lens itself. The function of the condensing system is to gather light efficiently from the source, while the projection lens projects an image of the object.

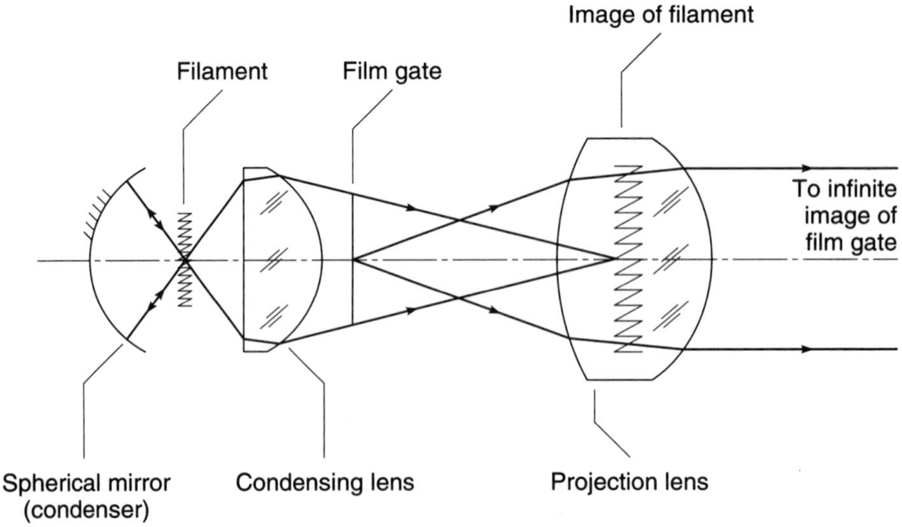

Figure 4.4 Classic projection scheme

For simplicity, imagine that the source emits light uniformly in all directions (in reality the emission pattern will be complex and usually presented as a polar diagram). Without the spheric mirror, all of the light emitted to the left in Figure 4.4 will be lost from the system and will not contribute to illuminance in the far field. The spheric mirror is used to capture this 'backwards light' in such a way as to enable the condenser lens to maintain an image of the source at the entrance pupil of the projection lens. A high screen luminance is then maintained.

A most convenient way of achieving this is to use the spheric mirror to form an image of the source beside itself.

Placing the source at the centre of curvature of the mirror forms an image of the source at the original source position. This is a result predicted from the paraxial '2*f*' rule, where a well-corrected image of unity magnification is formed when both object and image are placed at conjugate distances equivalent to twice the focal length of the lens. An alternative view is simply that the spheric surface is normal to rays from the source by definition and they are therefore returned along the same path.

In practice it is usual to have a source with an aspect ratio half that of the object to be projected. This is to enable the source image to be placed beside the source thus creating the full aspect ratio required (see Figure 4.5). Many tungsten–halogen filament projection lamps are manufactured with a filament aspect ratio half that of a 35-mm transparency for just this reason.

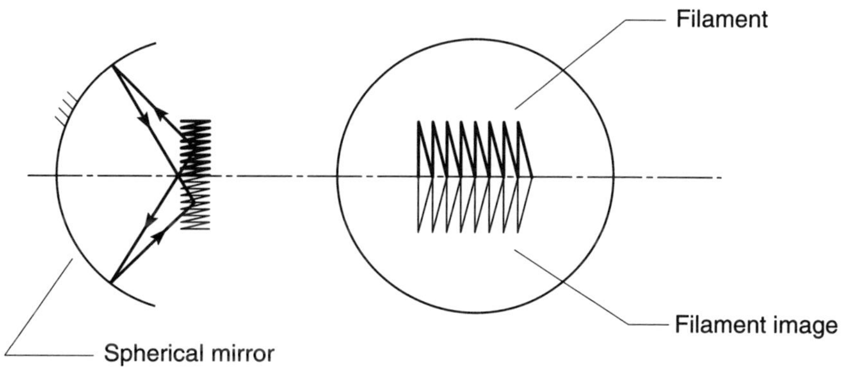

Figure 4.5 Filament aspect ratio

If the image of the source is placed directly over the source itself then the lamp filament willl obscure the returned light and limit the usefulness of the spheric mirror.

Parabolas

The parabolic mirror finds many applications as a collimator, from car headlamps through to off-axis interferometer components. Depending on where, how or for what the mirror will be used, its method of manufacture can vary enormously. This might be from a simple spinning process through to single-point diamond turning for high-quality infrared components.

One common application is in the reflecting telescope. If a large aperture is required, say for an astronomical telescope, then it soon becomes impractical to apply large refracting lenses, for considerations of mechanical stability (lenses can only be supported by the edge), glass quality, stria and bubbles, etc. In contrast, a large, front-surfaced mirror can be supported across its entire area and is relatively easier to fabricate.

Several different forms of reflecting telescope exist and two, the Newtonian and Cassegrainian, are illustrated in Figure 4.6. Both of these telescopes utilize concave paraboloidal primary mirrors.

The form of a parabola, with the vertex at the origin, is given by

$$Y^2 = 4fX$$

where f is the distance from vertex to focus (see Figure 4.7).

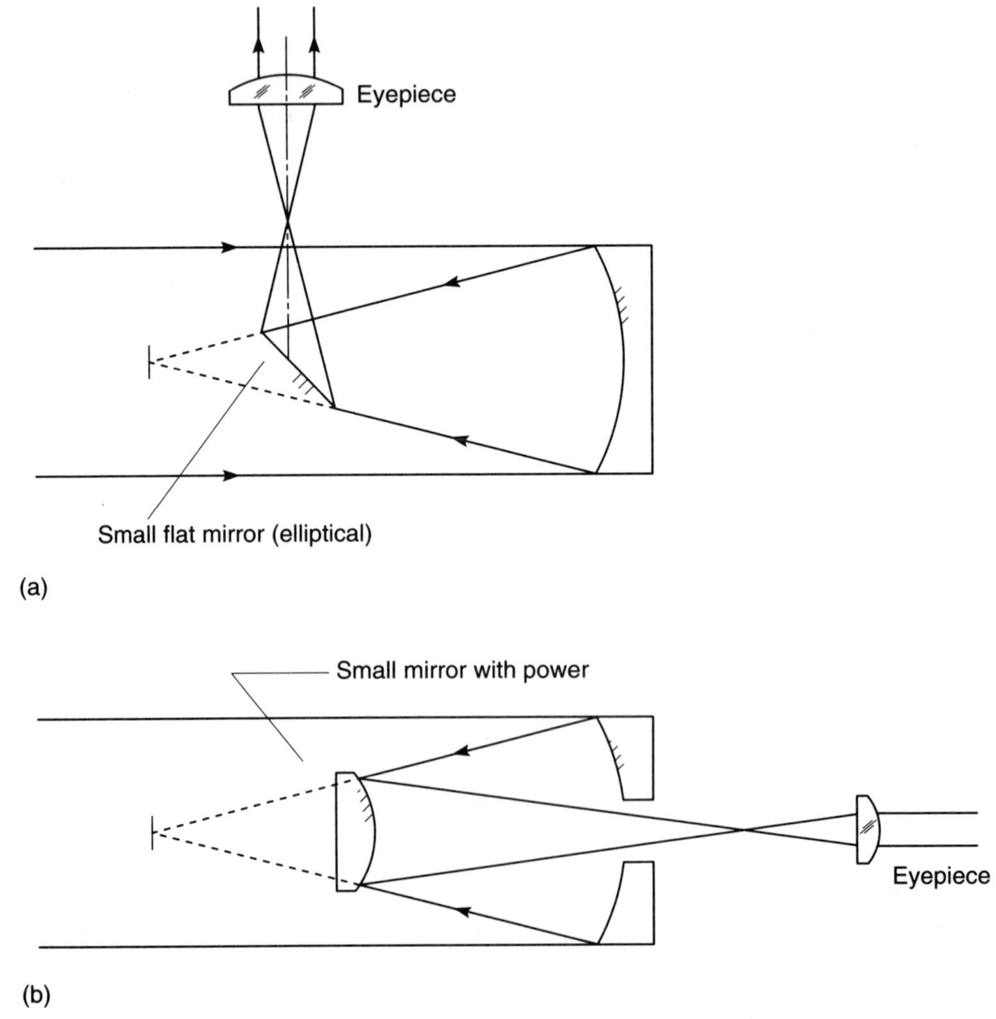

Figure 4.6 Reflecting telescopes: (a) Newtonian, (b) Cassegrainian

In the paraxial region spherical and parabolic mirrors are in form virtually indistinguishable (see Figure 4.8). Thus, for the paraxial region only, the relationship between object and image distances is the same.

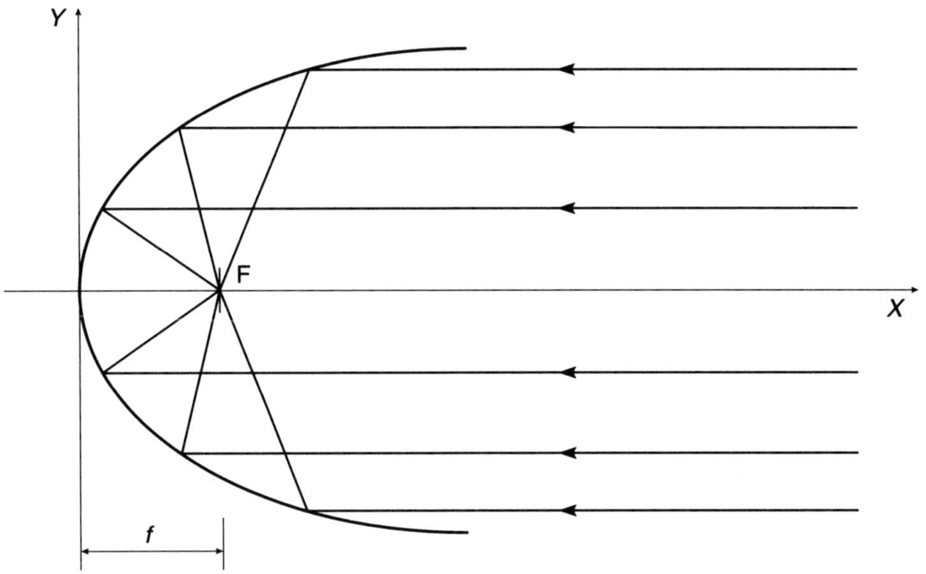

Figure 4.7 Parabola: $Y^2 = 4fX$

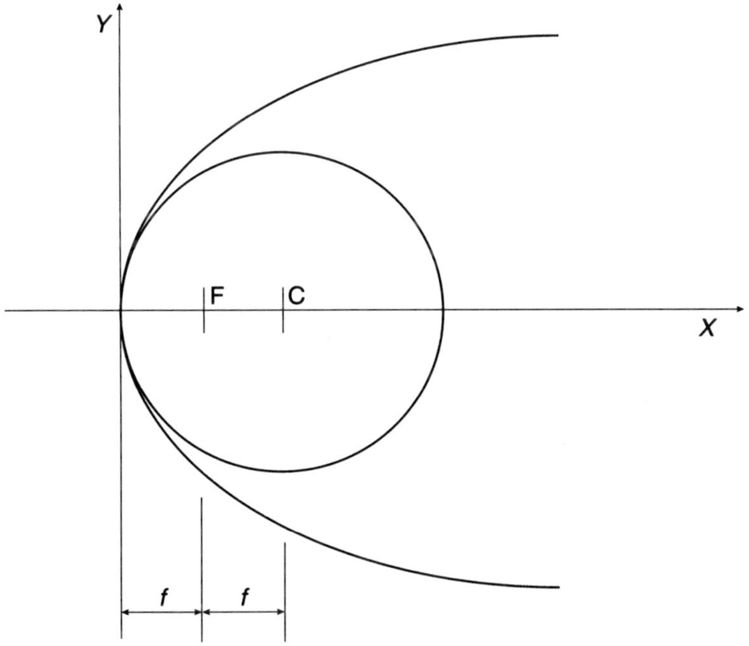

Figure 4.8 Parabolic vs. spherical reflector forms

It is interesting in passing to note that for a collimated pencil of rays in the paraxial region a parabolic mirror converts the ray bundle to a perfectly converging spherical shell.

As stated, the spherical imaging representation holds true only in the paraxial region. As the incident bundle of rays moves further out from the axis toward the limit of the physical aperture, the focal length of the parabola increases (see Figure 4.9(a)). This variation in focal length can be extreme for parabolas of high aperture. The increase in focal length is given by the following form, where F represents the distance from an off-axis point to the paraxial focal point.

$$F = f + x = f + Y^2/4f$$

Figure 4.9(a) shows that an off-axis parabola is a section taken from the margin of a full parabola, and illustrates clearly the increase in focal length towards the edge of a full parabola.

The imaging performance of a parabolic mirror falls very rapidly with field angle. As an example, large astronomical mirrors of the 1960s, while returning an excellent on-axis performance, suffer from coma with only a few seconds of arc of field. To some extent this can be cured through the application of a Schmidt corrector plate (see Appendix A), but the Schmidt corrector really comes into its own when used with a

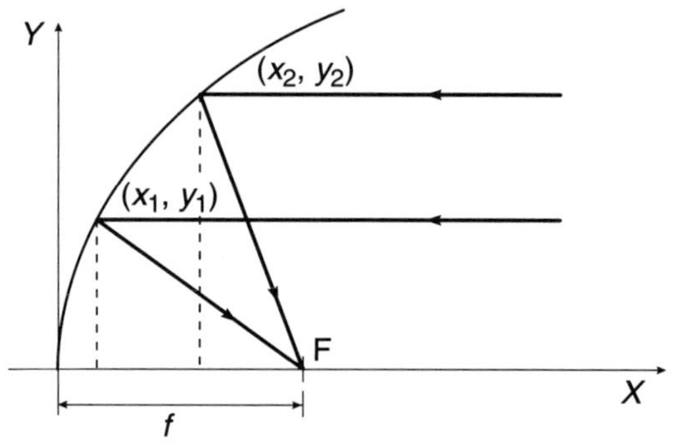

(a)

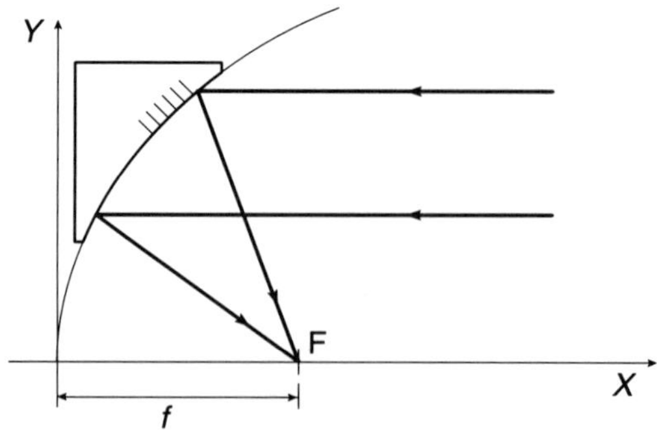

(b)

Figure 4.9 (a) Parabolic mirror: variation of focal length with aperture. (b) The off-axis parabolic mirror telescope having a spherical primary mirror, where the corrector can be placed at the centre of curvature. This arrangement essentially converts the system to have the characteristics of a parabolic primary mirror but corrected over all field angles. This arrangement is often called a Schmidt camera.

Anyone who wishes to pursue a discussion of the aberration characteristics of the parabolic mirror is directed to *Modern Optical Engineering* by Warren J. Smith (see Further reading).

As an example of how some of the characteristics of a parabolic mirror can be utilized we look briefly at an infrared, Fizeau-type interferometer for large-aperture optics (see the lasers section of Chapter 8 for a discussion of the application of interferometers).

Figure 4.10 shows schematically the layout of an unequal path laser interferometer of the Fizeau type, often used in the assessment of optical components. The laser beam from a CO_2 laser for operation at a wavelength of 10.6 µm is cleaned and collimated by a spatial filter that comprises a focusing lens and small pinhole. The size of the pinhole should be small enough to prevent any non-perfectly collimated light from the laser passing through the system, but large enough to not vignette the beam. The diameter of the pinhole is usually selected to be near the Airy disc diameter.

The cleaned and now diverging ray bundle is recollimated by a simple small lens, probably a singlet. The lens is small because infrared materials are expensive and a singlet should suffice because of the monochromatic source. A further singlet lens focuses the small-diameter collimated beam near a second steering mirror that directs

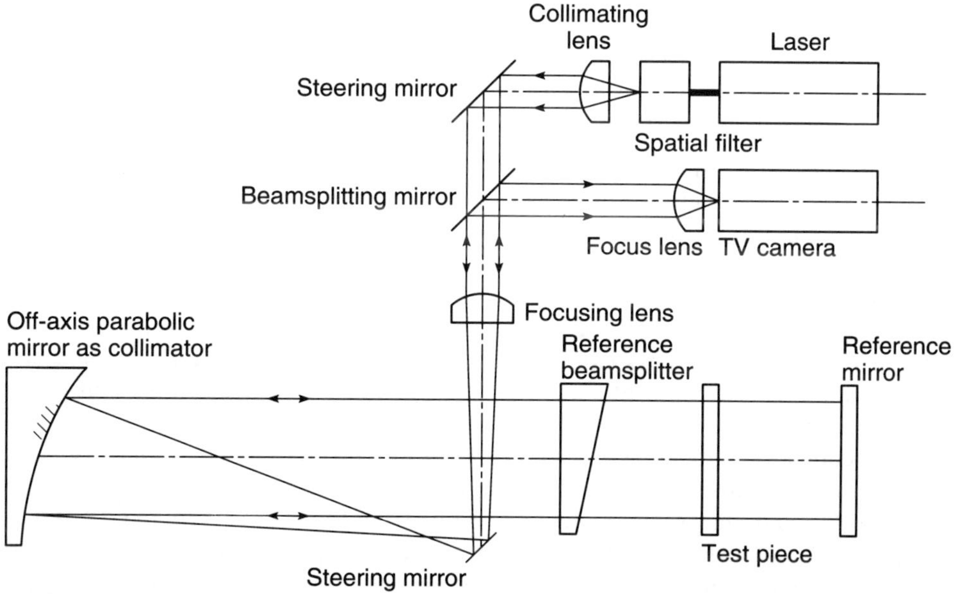

Figure 4.10 Fizeau interferometer with camera. The reference beamsplitter is wedged to prevent a reflection from the second surface interfering with the return beam

the now diverging ray bundle towards the off-axis parabolic collimator. The focus of this lens is near and not on the small steering mirror, so as to minimize beam disturbance due to dust, dirt or mirror imperfections.

Figure 4.10 shows clearly the advantage of an off-axis parabolic collimator. The off-axis mirror allows the construction of an unobscured, large-aperture collimated beam, whereas an axially produced beam would suffer from obscuration caused by the small steering mirror. The mirror allows large apertures to be created at a fraction of the cost of an equivalent refracting lens. This is particularly so when the mirror can be produced via the diamond turning process, which is most appropriate for infrared applications.

The highest-cost item in this particular layout would be the reference beamsplitter. This would be manufactured from a suitable infrared transmitting material, polished to interferometric standards and wedged to prevent the second reflection from the rear surface of the beamsplitter traversing the return path of the interferometer.

In this application the off-axis parabolic mirror has allowed the use of a relatively low-cost manufacturing process, and allowed also an unobscured collimation of the laser beam.

Ellipses

When a source or object is placed at the first focus of an elliptical mirror, a unity image of that source is produced at the second focus. This fundamental characteristic of the elliptical mirror makes it particularly useful for applications such as cinematographic projection, where the image of the source must be placed at or near the entrance pupil of the projection lens.

The elliptical mirror is a beam concentrator that allows the efficient collection of light from a small source and the reimaging of that source at a convenient location. The efficiency question can be divided to a number of issues, as follows:

- Through good utilization of the source polar output (intensity with respect to angle in 3D) most of the light emitted from the source can be collected and used. An alternative statement of this characteristic is that the reflector should subtend a large solid angle at the source. By way of an example, it will always pay in terms of collection efficiency to minimize the dimensions of the bulb/lampholder hole.
- With high reflectivity from the surface of the ellipse absorption losses are minimized. High reflectivity can be achieved with a number of different materials and coatings. These material combinations fall into a number of different groups, the most common of which are given below in the materials section.
- A high collection efficiency is in part achieved through good reimagery of the source. A highly aberrated image could, for example, result in the overfilling of a detector. Good imagery can be achieved only when the source is small compared to the physical dimensions of the reflector. A very rough rule of thumb is that the reflector throat should be at least ten to twenty times larger than the effective source diameter.

A feature of the elliptical mirror is that a point source placed at one focus is formed without spherical aberration at the other focus. This feature is used to good effect with Gregorian-type telescopes. The Gregorian telescope uses a parabolic mirror to form a well-corrected axial image of a distant star that is then relayed without spherical aberration to an eyepiece by an elliptical mirror (see Figure 4.11). The Gregorian telescope, unlike many astronomical telescopes, produces an erect image of the object.

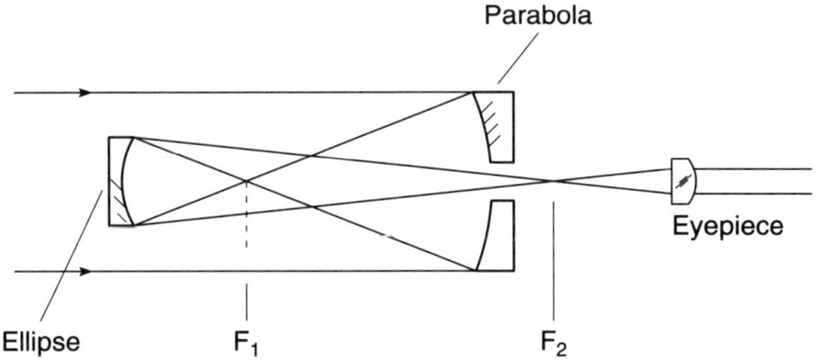

Figure 4.11 Gregorian telescope

An ellipse with centre at (x_0, y_0) and major axis parallel to the x-axis is defined as below.

- Length of major axis $= 2a$
- Length of minor axis $= 2b$
- Distance from centre to focus (F or F') is given by:

$$c^2 = a^2 - b^2$$

- Eccentricity is given by:

$$\varepsilon^2 = c^2/a^2 = (a^2 - b^2)/a^2$$

- Rectangular equation:

$$(x - x_0)^2/a^2 + (y - y_0)^2/b^2 = 1$$

Thus, if an ellipse has a major axis of 100 mm and a minor axis of 50 mm then we can calculate the separation of the foci and the reflector-pole-to-focus distance as follows:

$$2a = 100 \qquad a = 50$$
$$2b = 50 \qquad b = 25$$

The ellipse-centre-to-focus distance is given by:

$$\begin{aligned} c &= (a^2 - b^2)^{1/2} \\ &= (50^2 - 25^2)^{1/2} \\ &= 43.30\,\text{mm} \end{aligned}$$

Foci separation is given by $2c = 86.6$ mm, and the pole-to-foci distance is given by $a - c = 6.70$ mm.

Free form

The free-form reflector is one that is not rotationally symmetric and does not conform to notions of symmetrical aperture.

The free form can be a blending of two or more forms, such as parabolic in one axis and elliptical in the other, or can be completely non-parametric, i.e. defined by a number set only. Reflectors such as these pose a not inconsiderable design task, with the reflector surface having to be designed literally element by element. Their application is in very demanding lighting tasks where beam control and intensity are critical. One such area where these reflectors now find application is that of automotive lighting, particularly dipped-headlamps.

The traditional headlamp comprises a dual-filament lamp. One filament utilizes the whole of the reflector for the intense main beam and the other is shielded by an end-cap so as to use only part of the reflector to produce the dipped-beam pattern. This approach does not utilize all of the lamp's polar output in dipped mode and is therefore quite inefficient. Through the application of a free-form reflector the lamp end-cap can be dispensed with, and the dipped pattern formed through reflector shape only, with the attendant increase in beam power through high-polar output utilization.

This field of reflector design is still relatively new, and in essence no formulation can be usefully presented in a text such as this. The quality design of these reflectors should be tackled through CAD tools only.

Concentrators

A concentrator is a non-imaging reflector with the purpose of increasing flux at a given point in space. Applications are usually restricted to detector or solar-energy-collection systems.

Two such concentrator forms are the conical and compound parabolic, each rotationally symmetric for small detectors or of a trough-like form for long detectors or receivers; shown in Figure 4.12. The conical form is less efficient, owing to a requirement for a shallow cone angle. If the cone angle is large then rays entering the collector can be reversed in direction and escape the system after only a few reflections. The parabolic type of concentrator can, however, be designed to get the ray to the detector in only one or two reflections. The parabola is rotated about the detector and not necessarily its own axis.

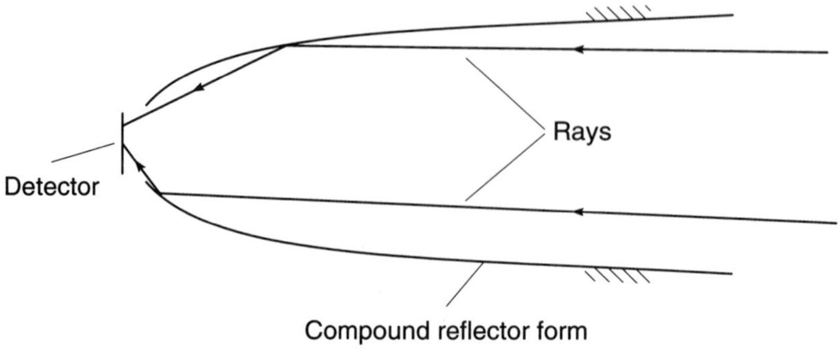

Figure 4.12 Compound parabolic concentrator

It is important to arrange for the ray to reach the detector with as few reflections as possible, thus minimizing losses due to reflection efficiency.

For a detailed discussion of concentrator technology you should read *The Optics of Non-Imaging Concentrators, Light and Solar Energy* by Welford and Winston (see Further reading).

Materials

Reflectors can be manufactured from a wide variety of materials, with a wide variety of different coatings and finishes. The section on optically flat mirrors has already presented some information with respect to flatness and quality requirements, but we turn now to the materials themselves and their suitability for different applications.

Reflector manufacturing technology can be separated into two groups, scientific and commercial. Definitions are as follows.

Scientific reflectors are those used in apparatus where the quality of reflection is important to maintain a good image or known spectral balance.

Commercial reflectors are those used for condensing or lighting purposes where quality of image is not quite so important.

There are of course some types of mirror that cross the boundaries and some that will not fit these definitions. A few of these types are discussed below under 'Others'.

Scientific mirrors

Scientific mirrors are commonly manufactured from a glass substrate with a high-reflectivity coating applied.

The glass blank is ground and polished to achieve the surface form and finish required by the application. Substrate materials are usually selected for low thermal expansion, stability and compliance with working techniques. A suitable material might be a borosilicate glass, such as Corning's Pyrex, that has a low expansion coefficient and is fairly readily worked.

The low thermal expansion coefficient is important if the mirror is, say, a spheric condenser used with a hot light source such as a small arc. The mirror will be robust toward the thermal shocks received during its lifetime.

The low-expansion glass can be finished with many different coatings. For the highest reflectivities, enhanced aluminium, silver or even gold can be used under vacuum deposition – the range here being from, say, 90+ per cent reflectivity from a low-grade aluminium up to 99+ per cent for gold. Coatings such as enhanced aluminium will require an overcoat of, say, silicon oxide (SiO_2) or magnesium fluoride (MgF_2).

More specialist coatings are available to produce hot or cold mirrors. Hot mirrors reflect hot, infrared, radiation, while transmitting the visible portion. The converse is true of a cold mirror that reflects the visible and transmits the infrared. These mirrors obviously have application for the control of thermal effects within projection and many other systems.

For very stable applications, mechanical and thermal, some large-aperture scientific mirrors may be manufactured from aluminium substrates with vacuum-deposited coatings. For the infrared these mirrors might be manufactured directly from the single-point diamond turning process.

For the most demanding applications, such as, say, high-power CO_2 laser beam handling for welding or cutting, mirrors might be manufactured from copper substrates with water-cooling channels. The surface might be diamond-turned, flat, spheric or aspheric, and subsequently coated with gold to ensure absolutely the highest reflectivity. At very high power densities even the smallest residual thermal absorption might be enough to distort the mirror (and therefore steer the beam where you don't want it) or even to destroy the mirror itself.

Commercial mirrors

Commercial mirrors or reflectors are very different to scientific mirrors in that price is usually the most important factor. Obviously, the cheaper a product can be made then ultimately the higher will be the profit margin. Such products will include lighting systems, luminaries or domestic lighting systems.

The mirror (or, more accurately, reflector) substrates are formed by mechanical processes such as sheet metal working, pressing, extrusion or spinning. Sometimes the surface will be left bare, while for others a brightening process may be applied.

In the electro-brightening process, applied commonly to spun reflectors, the super-pure aluminium surface is chemically polished, to remove micro-pits and scratches, and then clear-anodized to ensure long maintenance of the surface. A good finish from this process would result in a reflectivity of 60 per cent to 70 per cent.

Other finishes encountered include painted, for diffuse reflections, or heavily textured for control of beam diffusivity and colour.

Others

Plastic injection moulding, together with a low-temperature aluminium coating, perhaps with a lacquer coat between, is common for the mass production of fair-quality mirrors. Reflectors manufactured in this way would be suitable for, say, the production of torches, cycle lamps or similar mass market applications.

Investment in tooling can be expensive with several mould cavities engineered in each tool, but, provided the production run is long and the product cycle robust then the investment will be returned time and time again.

Reflectivities and surface finish can be good, indeed very good from new tooling, and complex shapes can be manufactured very easily, once the tool is made. The reflectors of course are very light in weight, but they are relatively mechanically and thermally unstable.

One final and very accurate method of forming reflectors for both scientific and commercial purposes is that of electro-forming. Here a highly accurate tool is formed, perhaps CNC-machined and polished with finishing by hand, on which is deposited a thin layer of the reflector material, say nickel. The nickel layer is then reinforced with subsequent depositions of material or even glass fibre laminations. The final product is then forced from the tool by mechanical means.

The material is deposited on the tool by electro-deposition – the migration of, say, nickel molecules in an electric potential. The reflector tool is immersed in a conducting medium. The very accurate inner surface can be subsequently coated with a material such as silver or gold if even higher reflectivities are required by the application. If a high-temperature vacuum deposition is required to finish the reflector then care must be exercised in the selection of the supporting material.

CHAPTER 5

Commercial components

Introduction

Once a specification has been finalized and you know what the lens has to do, what size it should be and how it should perform, you are left with the task of sourcing it. This means not only finding the lens but also finding it for the right cost and delivery.

There are, of course, two options. Either you can find (or ask someone else to find) the lens off the shelf, in which case you may have to compromise the specification, or you can get the lens custom-designed yourself.

Custom design will involve finding the right designer and the right manufacturing facility – right not just for the optical elements but for the mechanics, assembly and quality testing of the lens before you accept it for your own product or particular use.

Sourcing a lens off the shelf can be easy if your requirements are simple, or difficult if you have a specific high performer in mind. An off-the-shelf lens might be a biconvex singlet to improve a broadband radiometer, or it might be a precision semiconductor laser collimator with beam-shaping and correction optics. The off-the-shelf lens might be purchased from one of the large mail-order companies, or it might be a photographic camera lens, a microscope lens or a CCTV camera lens.

This section discusses some of the options available, some common sources of lenses and, when to go for a custom design.

Buyers' guides

A number of buyers' guides are available, all associated with a parent optics magazine or association. The common, and most useful ones are the following.

Photonics Buyers Guide

This is from the publishers of *Photonics Spectra* magazine and actually comes as a four-part set of guides (1994 edition) that comprises:

1. *Photonics Corporate Guide*
2. *Photonics Buyers Guide*
3. *Photonics Handbook*
4. *Photonics Dictionary*

Lasers and Optronics Buying Guide

Produced by the same people who publish *Lasers and Optronics* magazine, this has a companion volume, *The Technology and Industry Reference.* Again, it is a comprehensive guide as to who supplies what and where.

Laser Focus World Buyers Guide

This comes from the suppliers of *Laser Focus World.* Like the two guides above, it is a good, comprehensive guide to component and instrument sourcing.

The one drawback with all the guides just listed is that, being primarily American publications, they are heavily USA-biased. In the UK and Europe the two main sources of information are *Electro Optics* magazine, published in the UK, which contains a regular but small buyers' guide, and the relatively new *Optics and Lasers Europe* (OLE) magazine, which has yet to produce its own guide.

One other source of information, primarily for the UK and European markets is the *Physics World Buyers Guide* issued through the Institute of Physics.

Mail-order lenses

There are very many vendors willing to supply their own off-the-shelf lenses of many shapes, sizes and applications. Usually the vendor has a large, glossy catalogue listing and describing their wares. In some respects it will depend on geography, your personal biases and your budget as to whom you select to supply standard lenses.

Let us take a look at three of the major players in this field – with apologies in advance to those we miss out!

Melles Griot

The Melles Griot catalogue is a weighty tome that contains a wealth of information on the products supplied. The latter are well specified and well laid out to aid component selection. This product catalogue also contains useful background information.

The catalogue can be supplied on floppy disk in a suitable format so that many optical design programs can read the data and ray trace lenses directly from the catalogue. This is a very useful facility for laying out simple optical systems with proprietary components.

Melles Griot can supply coatings, components, lasers, laser accessories and detectors and optomechanical mounting systems. The optical component list includes items such as

- Singlets
- Doublets and triplets
- Cylindrical optics
- Condensers
- Prisms
- Filters and etalons

- Mirrors
- Windows and beamsplitters
- Laser optics
- Microscope optics
- Gradient index lenses (GRINs)

Melles Griot is a comprehensive, one-stop shop with representation worldwide. The company can also supply custom-designed optics.

Spindler & Hoyer

Like Melles Griot above, Spindler & Hoyer has a comprehensive list of optical components, also well specified, together with a comprehensive optomechanical mounting systems section.

Again, like Melles Griot, Spindler & Hoyer has a computer floppy disk catalogue that interfaces with optical design software. In this case the optical design software is Spindler & Hoyer's own and is available from them at low cost.

The company can supply a range of different light sources including low-voltage lamps, arc lamps, high-pressure lamps, spectral lamps and lasers. A range of optical instruments is available.

The range of components available includes the following:

- Single lenses
- Micro lenses
- Achromats
- Projection optics
- Microscope objectives
- Eyepieces
- Condensers
- Telescopes and laser beam expanders
- Prisms
- Flat optics
- Mirrors
- Laser optics
- Thin film coatings

Spindler & Hoyer can also supply custom-designed optics.

Ealing Electro-Optics

The Ealing catalogue contains some tutorial information, though it is not in the same league as Melles Griot in this respect. What Ealing Electro-Optics does offer is a very wide range of optical components, not yet on floppy disk. A wide range of products is available, which includes:

- Manual positioners
- Motorized positioners
- Component mounts
- Light sources

- Lasers and accessories
- Monochromators
- Optical benches
- Optical tables
- Optical instruments
- Optical components
- Filters
- Microscope components
- Books

The components are fairly well specified – at least well enough to enable a single-wave analyses of a combination of lenses for example.

Among the many other major suppliers are firms such as Oriel, Newport, Rolyn, Edmund Scientific, etc. Each supplier has a slightly different range of goods with slightly different specifications.

If you can't find what you want from one vendor then look at another. Custom design really is, for most applications, an expensive option and it pays to shop around the standard component vendors.

The other places from which to source standard lenses are the optics buyers' guides, usually associated with a parent magazine, and the trade press itself. The freely available optics magazines are usually littered with advertisements offering optics hardware. Take a look at

- *Photonics Spectra*
- *Laser Focus World*
- *Lasers and Optronics*
- *Optics and Lasers Europe (OLE)*
- *Electro Optics*

Happy hunting!

CCTV camera lenses

These lenses, together with the 35-mm camera lenses discussed below, represent good value for money in terms of performance. Unlike the photographic lens the CCTV lens tends to be a little more resilient to use outside the visible part of the spectrum. This is because most camera tubes and certainly those CCD-based cameras respond to the near infrared as well as to the visible.

If you are looking for a lens to work in the near infrared, perhaps up to a wavelength of 1 μm or so, then a CCTV lens might fit the requirement. There are very many vendors, some of the mail order warehouse type, and a great variety of lenses. The two most common fundamental types are the ⅔-inch and the 1-inch lenses. These lenses are of course for ⅔-inch and 1-inch camera tubes with their respective field, or image plane, sizes.

The general optics user will not require auto-iris, auto-focus and auto-tea-making – a standard lens will usually do. As with the 35-mm photography lens below, mass production has made these lenses a very cost-effective source of optical power.

35-mm camera lenses

The 35-mm still-photography camera lens usually gives high performance at low cost. But caution – modern photographic lenses are designed to a tight cost-versus-performance specification, and one way this is achieved is by correcting the lens only over the range to which the film is sensitive, namely the visible region. The antireflection coating will also be optimized for the visible. If you require near infrared performance then a photographic lens may prove unsatisfactory.

Another way in which the cost-versus-performance specification is met is through high-volume manufacture. The economy of scale plays a significant part in keeping the cost down: just compare the cost of the optical and mechanical design plus the manufacture of the one lens you require with walking into the local discount shop and spending £150.

The lens will have some characteristics to note, besides optimization for the visible only (approximately!) as already mentioned. The usable field will extend a little beyond a 35-mm negative size for a 35-mm lens, larger for other formats. This field will be circular of course! There will be a fixed mounting-flange-to-image-plane distance for all focus and zoom settings.

Despite these restrictions, however, in common with the CCTV lenses already discussed, the photographic lens offers good value.

Microscope lenses

Microscope lenses can be used for instruments and applications other than microscopes and microscopy. However, do be mindful that the performance will have been optimized specifically for the microscope conjugates, as described in Chapter 2, which are the parfocal distance and tube length. The microscope objective will almost certainly be inscribed with its magnification and numerical aperture for use at these defined conjugates.

Having noted the words of caution above, we may observe that microscope objectives are most useful lenses when applied to lasers. They can be applied as collimators and focusing elements for use in spatial filtering. The lower-power 'telescope' or long-working-distance objectives are particularly useful in some imaging applications. It is the monochromaticity and low divergence of the laser source that really allows the user to get away with this.

If you are going to use a microscope objective with a laser, do ensure that the laser wavelength corresponds to that of your objective – that is, the visible!

There are reflecting microscope objectives available, usually of a catadioptric design and usually of high power ($\times 30$ to $\times 50$ approximately) that can be used outside the visible.

Custom design

Custom design is not a trivial exercise. A qualified and reputable lens designer has to be found with the right experience to design your lens. For example, if your lens is to work in the infrared, has your selected designer actually worked with lenses in this waveband(?). This is a simple point but an important one.

Once the first design of the lens is available, the theoretical performance looks acceptable, the number of elements is about right to keep material costs down and it looks manufacturable, the next exercise is to tool-fit and tolerance the lens.

Tool fitting means taking the ideal lens as designed and fitting the individual lens curvatures to a manufacturer's existing stock of tools from which the lenses are to be made. This normally involves a modification of some lenses and air-gap thicknesses, possible a change in material and some curvature modifications. Normally no significant loss in performance will be predicted, and rarely is an improvement achieved! If a new tool is required then this adds to the cost of the manufacture, so it is generally wise to opt for tool fitting rather than a suite of new tools.

Following tool fitting comes a tolerancing exercise where all of the variables of lens thickness, air thickness, centration, tilt, displacement, curvature and sometimes lens material and hence refractive index are assessed. This exercise will ensure that no component has a specification that is too extreme within the manufacturing budget.

Your selected manufacturer then makes the individual components, probably cements some of these together and mounts them as groups, fits them all inside a lens barrel, paints Bloggs & Co. on the outside and presents the finished lens to you. Remember that detailed lens specification; you now have to test the lens to that specification to ensure that it performs as it should. We hope it will!

This rather long exercise (perhaps 10 to 12 weeks, or more if you want to put a coating on the lenses) will be costly and time-consuming. Make sure that you really do not have an off-the-shelf solution before moving to a custom design.

There are of course other reasons for a custom design, for example protection of an instrument design. If you require a special lens or optical component to achieve some aspect of your performance requirements and this special part is custom-manufactured on your behalf then some degree of market protection has been achieved.

CHAPTER 6

Measuring light

Introduction

In this chapter we discuss the optical spectrum in terms of a visible stimulus to the eye. We introduce the regions of interest to the optical worker outside the visible region of the electromagnetic spectrum, such as the infrared or ultraviolet. We also present some limited human eye physiology.

We explain the fundamental differences between radiometry and photometry, and discuss the question of when to use one measurement regime in preference to the other. Finally some basic measurement-of-light 'good housekeeping' rules are presented.

For the unwary and inexperienced, the measurement of light can often prove to be a minefield, littered as it is with strange photometric units such as the apostilb, the nit, the foot lambert or the talbot. So it is important to understand the distinction between photometric and radiometric measurements, the circumstances in which each applies and their relationship with one another. This understanding is especially important in using the correct units for a measurement or in a product specification. To assist your navigation through this chapter we begin with a crash course on photometric units and their radiometric equivalent. This information is summarized in conversion tables at the end of the chapter. Throughout the book photometric and radiometric units will be differentiated between by using a 'v' or an 'e' subscript respectively.

Since the human eye is the basis for all photometric measurement, the eye as a detector is closely studied. This includes the perception of colour and the various means for mathematically quantifying it.

Other types of detector are then dealt with, from silicon through ultraviolet to, finally, infrared detectors.

Photometry

Photometry is the measurement of light using the human eye as the reference detector, and consequently is only meaningful in measuring visible radiation. This limits photometric measurement to the wavelength range between 380 nm and 770 nm (deep violet to deep red). Radiometry of course has meaning throughout the electromagnetic spectrum, but is often not as useful in describing the visible spectrum.

The five most common measurement quantities are energy, power, flux density, intensity and brightness. Formal descriptions of each of these quantities are of less value than their relevance under real measurement conditions. Note, each of these quantities has a radiometric equivalent for which the formula, laws and advice apply equally to both measurement regimes.

Energy

lumen seconds ≡ joules

Photometric energy, Q_v, is measured in talbots or lumen seconds and is most often encountered in flashlamp or photographic applications. For example, a high-powered xenon flashlamp with a 600-joule input can produce over 30 000 talbots.

Luminous energy is also used to define the ASA rating or DIN index for a photographic film, as given by either

$$\text{ASA rating} = 5/E_c \tag{6.1}$$

or

$$\text{DIN rating} = (10 \log E_c) - 5 \tag{6.2}$$

where E_c is in lux seconds. Note that the film sensitivities refer to the luminous energy required to expose the film to a standard density when it is processed.

Power

lumens ≡ watts

Photometric power or luminous flux, Φ_v, is measured in lumens (lm); its radiometric equivalent is the watt (W).

The best means for measuring the output power of a light source is to enclose it in an integrating sphere. The inside of the sphere is coated in a highly reflective material such as titanium dioxide or barium oxide, and this enables it to collect the entire light output from the emitter. A calibrated detector measures the total power radiated and gives a direct reading in lumens.

Luminous intensity

lumens per steradian ≡ watts per steradian

The luminous intensity, I_v, of a source is a measure of its power per unit solid angle and is measured in candelas (cd), i.e. lumens per steradians.

$$I_v = \Phi_v/\Omega \tag{6.3}$$

With reference to Figure 6.1(a), the solid angle (Ω) can be calculated approximately from:

$$\Omega \approx A_d/d^2 \tag{6.4}$$

or, if the area is circularly symmetric about the measuring axis, as in Figure 6.1(b),

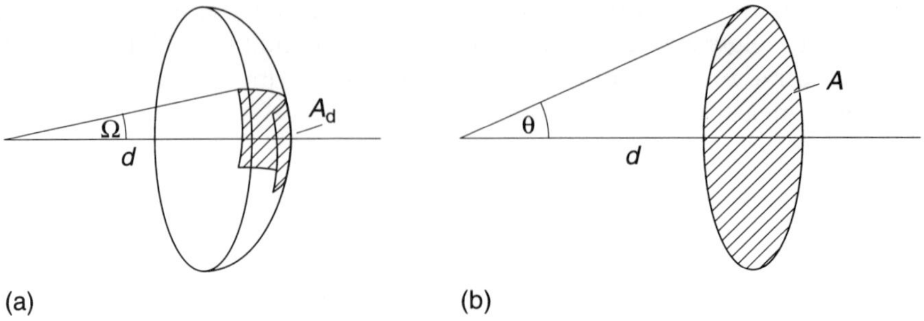

Figure 6.1 Definition of solid angle

then the solid angle is given by

$$\Omega = 2\pi\,(1 - \cos\theta) \qquad (6.5)$$

For a very small light source or a 'point' source that radiates in all directions into a sphere the solid angle is simply 4π steradians. If the source emits 1000 lm then the intensity is

$$I_v = 1000/4\pi = 79.6\,\text{cd}$$

Few light sources are really 'point sources' or radiate equally in all directions. More often the source has an area (A_s) and radiates preferentially in certain directions. Although the calculation of solid angle from Equation (6.2) or (6.3) is strictly limited to that area of a sphere subtended at a radius *d*, in practice any area can be used to give a good approximation, provided that the ratio of area to distance is greater than about 1:5.

Combining Equations (6.1) and (6.2) allows us to relate the luminous intensity to the illuminance (E_v) or power per unit area, hence

$$I_v = E_v d^2 \qquad (6.6)$$

This form of the intensity equation is very useful; if the source is small then a photometer can be placed 1 m from it and the lux reading converts directly into candelas. Alternatively, if the source is relatively large (10 to 200 mm), then the inverse-square law can be ensured by setting the photometer at a distance of $\sqrt{10}$ m. The lux reading need only be multiplied by ten to give the intensity in candelas.

For anything other than a point source the size of the emitting area has a bearing on the variation of intensity with angle. Most real light sources should be considered to be 'extended', and will to some approximation obey Lambert's law. Lambert's law relates the intensity of an extended source to the angle at which it is viewed:

$$I_\theta = I_0 \cos\theta \qquad (6.7)$$

Flux density

lux ≡ lumens per square metre ≡ watts per square metre

Flux density, whether illuminance, E_v, or exitance, M_v, is a measurement of the luminous power per unit area. For illuminance the area is that of the receiver, A_d, while for exitance it is the area of the emitter, A_s.

Illuminance

Illuminance can be calculated in a number of ways:

$$E_v = \Phi_v / A_d \qquad (6.8)$$

Alternatively, Equation (6.4) could be rewritten to show that for a point source radiating into 4π steradians the illuminance is inversely proportional to the square of the distance from the source, i.e. the inverse-square law applies:

$$E_v = \Phi_v / 4\pi d^2 \qquad (6.9)$$

It is always worth while checking that the inverse-square law applies at the measurement position. Sometimes stray light, background lighting or multiple reflections can affect the reading. In addition, a near-field-intensity distribution, which can mask the inverse-square law close to the source, can cause measurement discrepancies. Doubling the distance from the source to the detector should reduce the signal by a quarter. If it doesn't, try masking the detector to limit the stray light which can reach it, or repeat the test at double the original distance.

Cosine laws

The relationship between intensity and illuminance stated in Equation (6.6) can now be expanded to a number of important measurement cases illustrated in Figure 6.2 for a Lambertian source. The formulae that result are often referred to as applications of the 'cosine law' of photometry or radiometry.

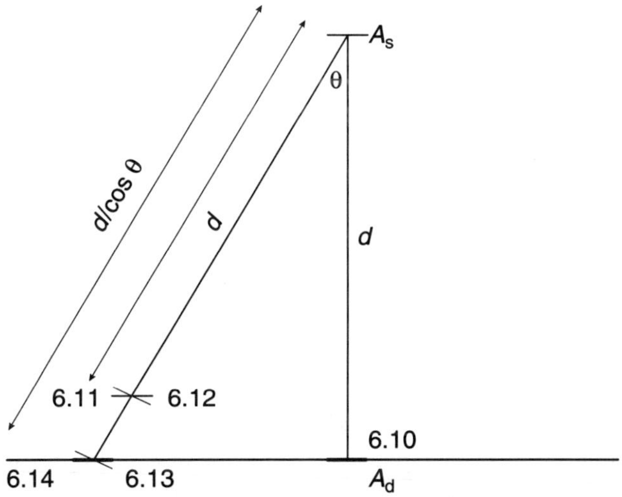

Figure 6.2 Cosine laws

The power at each position is derived from the luminance (L_v) by using the following formulae:

1. Light straight onto detector:

$$L_v A_s A_d / d^2 \tag{6.10}$$

2. Cosine term, owing to Lambert's law:

$$L_v A_s A_d \cos\theta / d^2 \tag{6.11}$$

3. Additional cosine term, owing to the projected area of the detector:

$$L_v A_s A_d \cos^2\theta / d^2 \tag{6.12}$$

4. As Equation (6.11), but with an extra cosine-squared term, owing to the increased distance from the source:

$$L_v A_s A_d \cos^3\theta / d^2 \tag{6.13}$$

5. A combination of conditions 3 and 4:

$$L_v A_s A_d \cos^4\theta / d^2 \tag{6.14}$$

In practice, this last case means that a detector such as a linear CCD should show a $\cos^4$ ('cos-fourth') variation in received signal between its centre pixel and the one at the edge. Note, too, that Equations (6.11) to (6.14) are reduced by a cosine term if the source, such as a point source, does not obey Lambert's law.

Although often quoted as $\cos^4$, the reality is often more complicated and the factor can vary from 2.5 to 5.

The impact of the $\cos^4$ variation is most keenly felt when it is necessary to illuminate a surface evenly at angles beyond 15°. This is a critical requirement in photocopiers, where the evenness of illumination determines the copy quality. To solve the $\cos^4$ problem the stop of the imaging lens is often profiled to reduce the amount of light through the centre of the lens, while leaving the edges unaffected.

Illuminance and lenses

Figure 6.3 shows a lens collecting light from a source and imaging it onto a detector at its focus. The collection area is now the area of the lens A_l, so the power falling on the detector is given by

$$\Phi_v = L_v A_s A_l \cos^4\theta / d^2 \tag{6.15}$$

To calculate the illuminance at the detector we know that $E_v = \Phi_v / A_d$, or:

$$E_v = \frac{L_v A_s A_l}{A_d d^2} \cos^4\theta \tag{6.16}$$

Assuming that area A_s now refers only to the area of the source which is imaged onto the detector, then:

$$A_s / d^2 = A_d / f^2 \tag{6.17}$$

This allows us to write an expression for the illuminance in the focal plane of the lens in terms of the lens parameters:

$$E_v = L_v A_l \cos^4\theta / f^2 \tag{6.18}$$

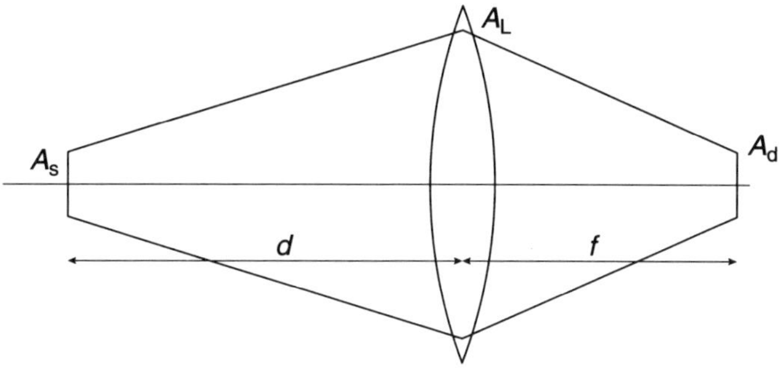

Figure 6.3 Illuminance at the focal plane

or alternatively for a circular lens:

$$E_v = L_v \pi \cos^4 \theta/(\text{f-number squared}) \tag{6.19}$$

Measuring illuminance

A photometer or light meter can be used to characterize any visible light source; the reading will usually be in lux (lm/m^2), and it is tempting to simply divide by the area of the detector and calculate the luminous power. This is valid only if the detector area is entirely and evenly filled. Even so, the power reading is only relevant for that particular position, and it can be misleading, especially in projection or display applications when the overall efficiency of the optical system is important. In such a case it may be better to measure the 'screen lumens'.

Screen lumens

When a projector creates an image on a screen, the central region is usually the brightest and the corners are darker. To make a realistic measurement of the power on the screen, flux density is measured at the centre and at four other points, each 70 per cent along the diagonal from the centre to the edge of the image. The five readings are averaged and then divided by the area of the image to give the 'screen lumens' value. This method is commonly used in the projection and display industry, and is shown graphically in Figure 6.4.

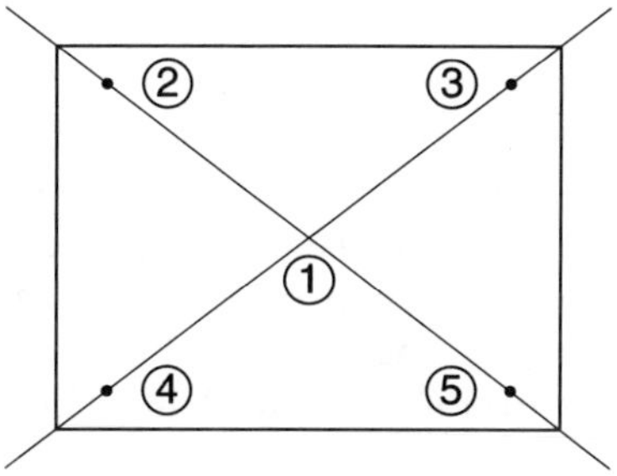

Figure 6.4 Measuring screen lumens

Lighting design lumens

Nearly all light sources have an ageing characteristic, which results in the output power decreasing with time. For a lighting designer trying to predict a realistic lighting level for a new luminaire it is advisable to base calculations on the output of a burnt-in lamp. This has led to the concept of lighting design lumens (LDL), which express the total luminous output of the lamp after a set time, usually 2000 hours for a fluorescent lamp and 100 hours for an arc lamp.

Illuminance conversion factors

Apart from the lux, there are several other units used to measure illuminance, and although included here they are seldom used (Table 6.1).

Table 6.1 Former units of illuminance

Unit	*Definitive value*
foot candle	1 lumen foot^{-2}
phot	1 lumen centimetre^{-2}
milliphot	1 lumen (1000 centimetres^{-2})

Conversion factors are given in Table 6.2.

Table 6.2 Conversion factors

	lux	*foot candle*	*phot*	*milliphot*
lux	1.0	10.76	10 000	10
footcandle	0.0929	1.0	929	0.929
phot	0.0001	0.00108	1.0	0.001
milliphot	0.1	1.076	1 000	1.0

Exitance

Although a lambertian source is one which radiates into 2π steradians, its exitance (M_v) is not 2π multiplied by its brightness (L_v). The reason for this becomes apparent when Figure 6.5 is studied. We already know that the intensity of the source is given by Lambert's law as $I_0 \cos\theta = L_v A \cos\theta$. Now the total power radiated into the hemisphere will be the product of the intensity and the solid angle. The area of the incremental ring on the surface of the hemisphere is $2\pi R \sin\theta \, R \, d\theta$, so its solid angle is $2\pi R^2 \sin\theta \, d\theta / R^2 = 2\pi \sin\theta \, d\theta$. The incremental power dP is therefore

$$dP = 2\pi L_v A \cos\theta \sin\theta \, d\theta \tag{6.20}$$

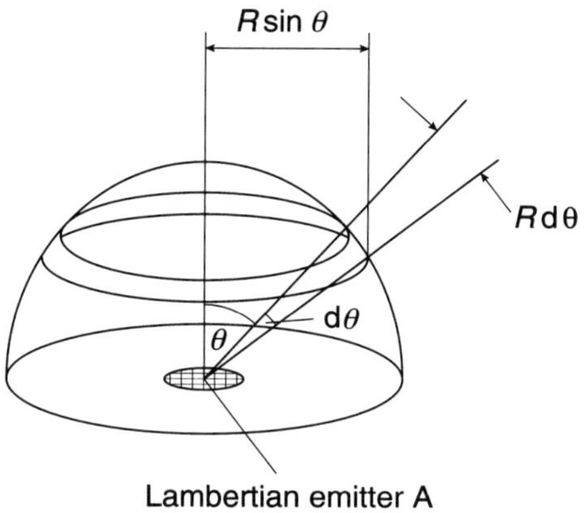

Figure 6.5 Lambertian source radiating into a hemisphere

The total output power is obtained by integrating this expression from 0 to $\pi/2$, and the exitance is simply P/A, so

$$M_v = \int_0^{\pi/2} 2\pi L_v \cos\theta \sin\theta \, d\theta = \pi L_v \tag{6.21}$$

This factor of π is already included in such units as the apostilb, footlambert and lambert.

Brightness

Lumens per square metre × steradians ≡ watts per square metre × steradians

Brightness or luminance, L_v, is a measure of the output intensity of a light source per unit area and its unit of measurement is the candela per square metre, or nit. To calculate the luminance of a source, use

$$L_v = I_v/A_s \tag{6.22}$$

or, for a point source

$$L_v = \Phi_v/A_s 4\pi \tag{6.23}$$

For an extended source obeying Lambert's law the luminance is constant over all angles. This is because although the intensity varies as a function of $\cos\theta$, the projected area of the source is also reduced by a $\cos\theta$ term. These two factors cancel each other out, giving this important result.

The combination of area and intensity means that it is never possible to increase the brightness of a light source, if both the object and the image are in the same medium. Although it might be possible to create a smaller image of the original light source using a lens, it would be possible only at the expense of collecting less light from the original object. This is really just an alternative statement of the Lagrange invariant, but it continually features as the limiting factor of many optical systems. Remember, if a high-efficiency optical system is required then use the brightest possible source to begin with. In some cases it is even better to use a lower-power source, with higher brightness, than to select a high-power lamp whose brightness is reduced because the source size is larger.

Luminance conversion factors

Although the nit is the most common unit of luminance, others are used in certain applications. For example, the foot lambert is often quoted in cinema or projection TV applications where the screen is evenly diffusing and the π-term is applicable. Conversion factors are given in Table 6.3.

The human eye

As the human eye has already been referred to as the yardstick for photometric measurement it is timely to discuss it in more detail.

Rods and cones

The light-sensing area in the human eye, called the retina, has two kinds of receptors, cones and rods.

Cones operate in light-adapted conditions and can sense colour; they are responsible for photopic vision. Light adaption is deemed to be at a luminance greater than 3 cd/m^2. This is slightly less than the typical luminance at sunset on an overcast day. Generally a healthy eye can become light-adapted after about 3 minutes.

Rods are for dark-adapted (scotopic) vision, but have no colour sensitivity. Dark adaption occurs below 30 μcd/m^2, which is darker than a moonless overcast night. Although the time taken to become dark-adapted depends on the initial luminance, full adaption can take up to 45 minutes.

Cones and rods are not evenly distributed within the retina. The fovea, which is the central area around the optic nerve, is composed only of cones, while further out at the periphery of the retina rods are more numerous. This is why we are more sensitive to movements at the edge of our field of view at night.

The dynamic range of the human is eye is enormous, ranging from 5 to 10 photons at the lowest level to millions of photons at the other extreme.

Photopic response

The relative sensitivity curve for the eye is shown graphically in Figure 6.6 and is listed in the $y(\lambda)$ column of Table 6.10 (p. 136). This is also the relative photopic response $V(\lambda)$, or daylight sensitivity of the eye.

Peak sensitivity in the green at 555 nm is an order of magnitude greater than the blue and about three times the red. The shape of the photopic response curve is the reason why the default-weighting factors in MTF calculations are 1.0 at 546 nm but 0.5 at 486 and 635 nm. For an optical system operating in visible light it is nearly always best to specify the nominal performance at a green wavelength.

Table 6.3 Luminance conversion factors

Unit	*Defining value*	*nit* *nt*	*apostilb* *asb*	*candela per square foot* *cd ft^{-2}*	*footlambert* *FL*	*stilb* *sb*	*lambert* *L*
nit	1 cd m^{-2}	1.0	3.141 6	0.092 9	0.291 9	0.000 1	0.000 314 16
apostilb	1 cd m^{-2} π^{-1}	0.031 83	1.0	0.029 57	0.092 9	3.183×10^{-5}	0.000 1
candela per square foot	1 cd ft^{-2}	10.76	33.82	1.0	3.141 6	0.001 076 4	0.003 382
footlambert	1 cd ft^{-2} π^{-1}	3.426	10.764	0.318 3	1.0	$3.426\,4 \times 10^{-4}$	0.001 076
stilb	1 cd cm^{-2}	10 000	$3.141\,6 \times 10^{-4}$	929	2 919	1.0	3.141 6
lambert	1 cd cm^{-2} π^{-1}	3183	10 000	295.7	929	0.318 3	1.0

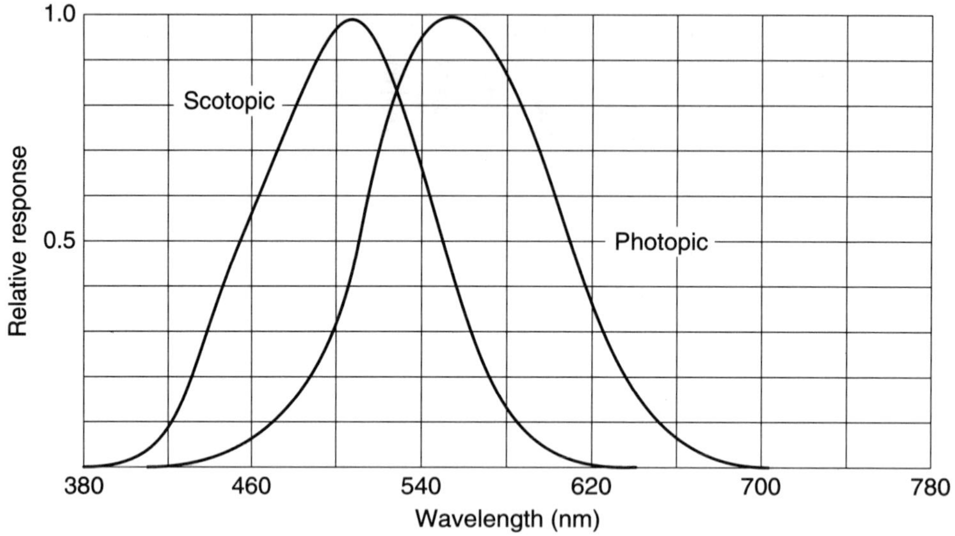

Figure 6.6 Photopic and scotopic relative responses

It is sometimes useful to know the relative amounts of light within a particular colour band. Table 6.4 gives this information for a source, which has a constant output through the visible spectrum (xenon arcs are close to this) and so is effectively the composition of the photopic response. For comparison, a light source with a blackbody temperature of 2856 K is included ('Illuminant A').

Table 6.4 Amounts of light within colour bands

Colour band *λ (nm)*	*Blue* *400 to 490*	*Green* *500 to 570*	*Yellow* *570 to 750*	*Red* *600 to 750*
Xenon arc	5%	56%	42%	19%
Illuminant A	2%	48%	53%	27%

Scotopic response

For night vision, where the rods are dominant, the spectral sensitivity changes to the scotopic characteristic (also shown in Figure 6.6). Peak response is now at 510 nm and the relative sensitivity of the eye has shifted to make it more responsive to blue and less sensitive to red.

Mesopic response

In between the extremes of scotopic and photopic vision the eye behaves in an intermediate fashion, termed the mesopic response. As the luminance reduces from 3 cd/m^2 to 30 μcd/m^2 the peak of the sensitivity curve gradually moves from 555 nm down towards 510 nm. As the light level is reduced the cones become less effective and the sensitivity becomes dominated by the rods. As luminance falls, the increase in light sensitivity of the mesopic eye is accompanied by a reduction in colour perception.

Spectral luminous efficacy

There comes a time when it is necessary to convert between photometric and radiometric units. These two measurements are related by the spectral luminous efficacy $K(\lambda)$, measured in lumens per watt and defined by

$$K(\lambda) = \Phi_{v\lambda}/\Phi_{e\lambda} \tag{6.24}$$

At the peak of the photopic response, at 555 nm, the luminous efficacy is 683 lm/W, so the relative spectral efficiency, $V(\lambda)$, is related to $K(\lambda)$ by

$$K(\lambda) = 683V(\lambda) \tag{6.25}$$

Spectral luminous efficacy for scotopic vision $K'(\lambda)$, is related to the scotopic spectral luminous efficiency $V'(\lambda)$ by:

$$K'(\lambda) = 1700V'(\lambda) \tag{6.26}$$

where 1700 lm/W is the maximum efficacy at 510 nm.

In older text books these values are sometimes quoted as 673 lm/W and 1725 lm/W, but the new values are now accepted standards.

Obviously, monochromatic sources such as lasers are easily assigned an efficacy figure, but sources with finite spectral width have to be integrated over the visible spectrum. Normally the spectral radiant power $\Phi_{e\lambda}$ of a source is known, so the efficacy can be calculated using

$$K(\lambda) = 683 \int_0^\infty V(\lambda)\Phi_{e\lambda}\,d\lambda \Big/ \int_0^\infty \Phi_{e\lambda}\,d\lambda \tag{6.27}$$

For blackbody sources only, this equation can be written as

$$K(\lambda) = 683 \int_0^\infty V(\lambda)\cdot M(\lambda)\,d\lambda \Big/ \sigma T^4 \tag{6.28}$$

were σ is the Stefan–Boltzmann constant and T is the temperature. The efficacy of some common sources are listed in Table 6.5.

Table 6.5 Luminous efficacy

	Peak wavelength	*Efficacy lm/W*
He–Ne laser	633 nm	148
Nd–YAG (frequency doubled)	530 nm	589
He–Cd laser	440 nm	15.7
Green LED	557 nm	625
Red LED	635 nm	130
Blue LED	467 nm	165
Xenon arc	800 nm	85
Filament lamp (2856 K)	1.0 μm	16

Sensing colour

The most useful model for understanding how the eye senses colour comes from the trichromatic theory. Here, the retina of the eye is divided into individual colour-sensing elements, rather like a colour CCD array. Each element is sensitive to specific wavelength ranges – broadly red, green and blue; these are the primary colours, but they do overlap considerably. A single colour is sensed from the combination of these three primary signals. Differences between individual observers and the limitations of the eye as a measuring instrument mean that we are usually better at colour matching than at defining an exact colour.

Colour perception is influenced by three distinct quantities: luminance, hue and saturation. Luminance or brightness is the most important, and the perceived shade of a colour can be altered by the brightness of the scene in which it appears. Next in importance is hue, which is concerned with colour distinction, for example the degree of redness, greenness or blueness of a colour. Saturation or chroma is concerned with the purity of a colour, effectively the amount of white that it contains.

Colour measurement and chromaticity coordinates

The theory that any colour can be defined by the combination of three primaries has lead to the most common means of defining and measuring colour. In 1931 the CIE (see Appendix A) produced standard colour-matching functions to define three supersaturated primary colours which can be combined without the need for subtraction. These functions, $x(\lambda)$, $y(\lambda)$, $z(\lambda)$, are shown in Figure 6.7 and tabulated in Table 6.10 and CIE Publication 18.2 (see Further reading).

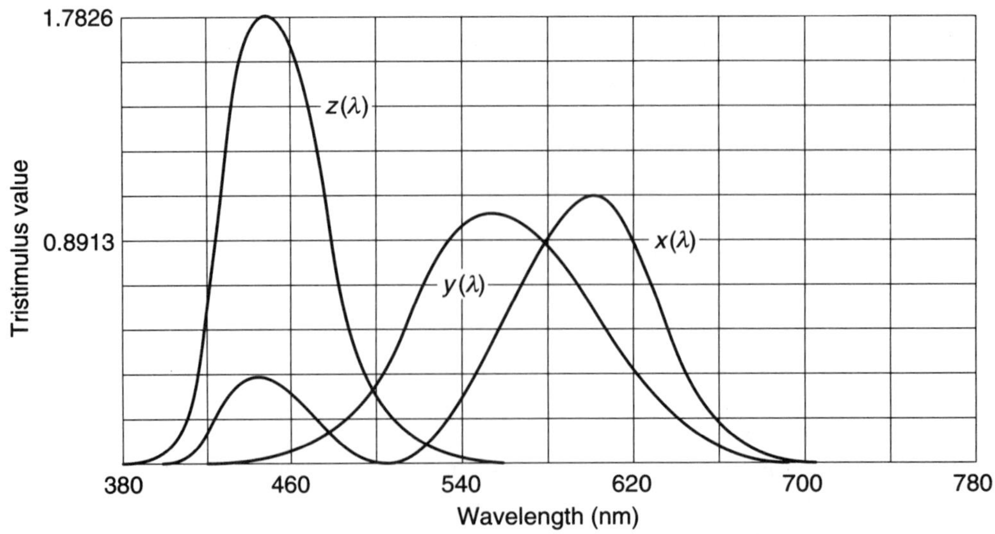

Figure 6.7 Tristimulus values for the standard colorimetric observer

Although these colours are not physically attainable, the y-curve has been chosen to match the normalized photopic response of the eye $V(\lambda)$ (see CIE Publication 18.2 – listed in Further reading). In doing this, luminance information is directly supplied to the colour-matching function. The other two curves are concerned only with chrominance.

Chromaticity coordinates

The colour-matching functions are each combined with a spectral response to generate the tristimulus values X, Y and Z. Chromaticity coordinates x, y and z can then be calculated from ratios of the tristimulus values:

$$\begin{aligned} x &= X/(X+Y+Z) \\ y &= Y/(X+Y+Z) \\ z &= Z/(X+Y+Z) \end{aligned} \tag{6.29}$$

As $x+y+z=1$, it is usual to specify only the x- and y-coordinates.

If the colour is not monochromatic, a summation process must be used to determine the tristimulus values over the wavelength range 380 to 760 nm:

$$\begin{aligned} X &= \int \phi(\lambda)\cdot x(\lambda)\,\mathrm{d}\lambda \\ Y &= \int \phi(\lambda)\cdot y(\lambda)\,\mathrm{d}\lambda \\ Z &= \int \phi(\lambda)\cdot z(\lambda)\,\mathrm{d}\lambda \end{aligned} \tag{6.30}$$

Chromaticity coordinates are often plotted on the chromaticity diagram shown in Figure 6.8.

Uniform chromaticity diagrams

Although the CIE 1931 coordinates are the most commonly used, they do suffer from the disadvantage of distorting the apparent sensitivity of the eye to different colours. In an attempt to overcome this, uniform chromaticity diagrams have been defined. In 1960 the CIE produced the first uniform chromaticity standard using u and v. These units are now less commonly used and have been replaced by the 1976 CIE u', v' diagram (see CIE Publication 17.4 – listed in Further reading), where:

$$\begin{aligned} u' &= 4X/(X+15Y+3Z) \\ v' &= 9Y/(X+15Y+3Z) \end{aligned} \tag{6.31}$$

1976 CIE coordinates are plotted in Figure 6.9.

Although the 1960 u, v are mostly supplanted by u', v' they are used in the calculation of correlated colour temperature (see Chapter 8). The two coordinate systems are related by

$$u = u' \qquad \text{and} \qquad 1.5v = v' \tag{6.32}$$

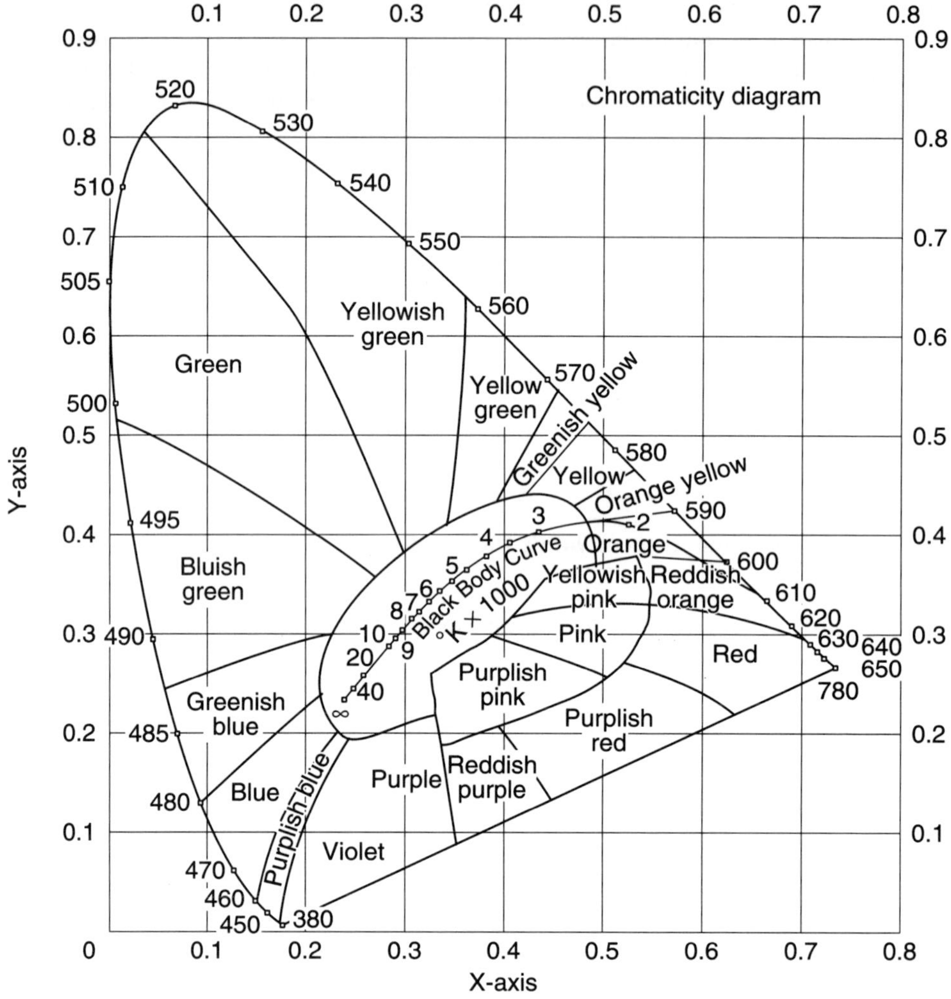

Figure 6.8 A CIE 1931 chromaticity diagram

By way of example, some chromaticity coordinates for typical filter colours are listed in Table 6.6.

Table 6.6 Chromaticity coordinates

	Blue	*Green*	*Yellow*	*Orange*	*Red*
x	0.15	0.333	0.512	0.598	0.6645
y	0.0523	0.625	0.48	0.3965	0.29
u'	0.18	0.135	0.265	0.364	0.5174
v'	0.1414	0.572	0.558	0.544	0.506

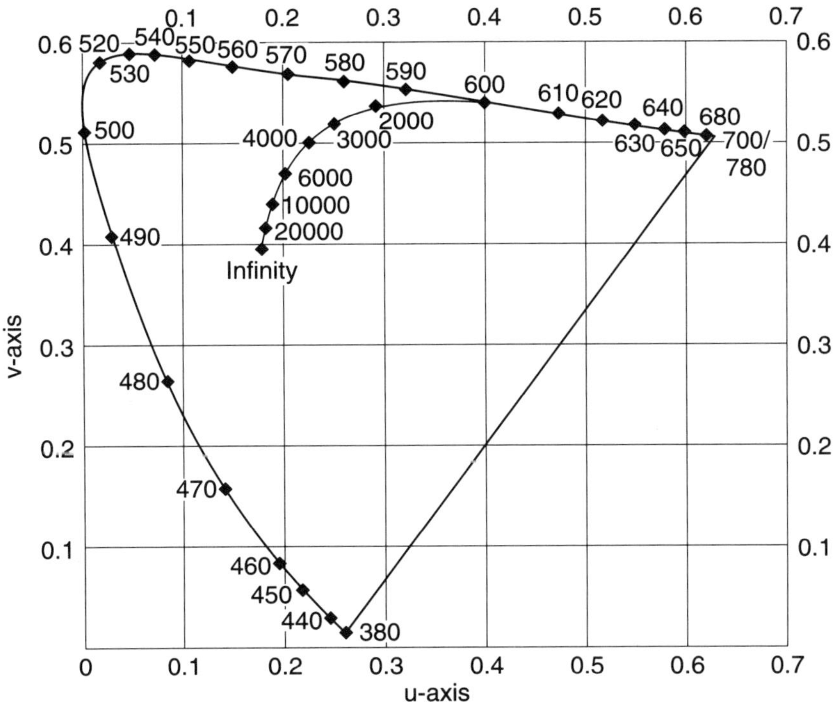

Figure 6.9 CIE 1976 UCS chromaticity diagram

Colour differences

Chromaticity diagrams can only represent colour differences at the same luminance. In reality, different colours can have different luminances and this can change our perception of them. To overcome this problem, a 3D colour space is required. Two alternatives have been created by the CIE: CIELUV, which is most applicable to TV and movie projection, and CIELAB, which is used extensively by the dyes and colouring industries. The methods for calculating L^*, a^*, b^* and L^*, u^*, v^*, shown in Appendix C, are slightly tortuous and do not lend themselves to an intuitive understanding of colour unless one is using them on a regular and specific basis.

Detectors

As mentioned elsewhere in this chapter, the eye is the human photoreceptor. It is the visible-light detector that has final authority on questions of brightness and colour perception. It is a remarkable performer with high resolution, large dynamic range and, together with its processors, some sophisticated pattern and motion recognition capabilities. To all of this we can add a long and reliable life during which little maintenance is required. However, in science and technology we need detectors that can respond across the visible, ultraviolet and infrared spectra for a variety of different tasks such as image recording, data communication, measurement and much more.

We encounter optoelectronic devices routinely during everyday life. Telephone conversations are conducted over long distances using fibre optic cables with laser diode sources and fast detectors. Our music is recorded and played back through laser and compact-disc technology using fast detectors for signal processing and automatic focus tracking. Supermarket doors open automatically and security lights follow us around a dark building. The optical bar code reader is at many point-of-sale counters. Complex and sophisticated optical detection systems are all around us.

This section discusses some of the optical detectors commonly encountered for the ultraviolet, visible and infrared spectra. The several subsections cover detectors of single-element type as well as some principles, area detectors, some examples and rules for good housekeeping. This is not an exhaustive survey of all the detectors available on the marketplace, but an introduction to some of the concepts and types that will be encountered by most electro-optic engineers.

Detection by a single element

This discussion is much 'biased' toward the visible spectrum and the ubiquitous PIN diode. However, ultraviolet and infrared detectors are introduced and a few salient points discussed. To cover the subject of detectors in detail would require a very large volume and is not the aim here. Further information with respect to detector physics and practicalities can be obtained from *Opto-Electronics: an introduction* by Chaimowicz (see Further reading).

This section starts with the ultraviolet and works through to the infrared. The electromagnetic spectrum is defined for our purposes according to Figure 6.10.

Ultraviolet

In general terms, the ultraviolet (UV) begins in the blue of the visible and decreases in wavelength to the X-ray region. However, the optical worker usually considers the UV to lie between about 100 nm and 400 nm.

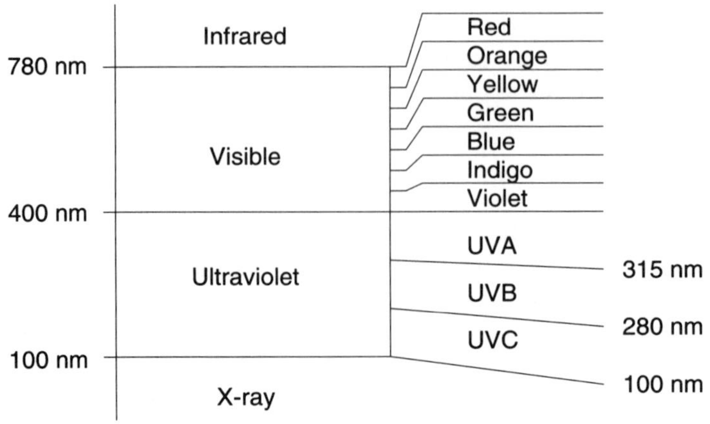

Figure 6.10 The electromagnetic spectrum

The CIE has subdivided the UV to three regions (CIE Publication 17.4):

- UVA: 400 nm to 315 nm.
- UVB: 315 nm to 280 nm.
- UVC: 280 nm to 100 nm.

One of the most important uses of the detection of UV radiation is in the field of health and safety. Without going into detail, we may point out here that the effects of exposure to UV radiation include skin burns and skin cancer. Treat UV with care!

The silicon PIN photodiode, discussed below, can be doped appropriately to give a response in the ultraviolet. However, the shortest wavelength at which these devices will return a usable signal is usually around 250 nm, where a responsivity of about 0.04 A/W might be achieved. This must be compared with the performance of a more traditional silicon photodiode, whose response will cease below about 400 nm.

For shorter-wavelength ultraviolet detection various physical constraints become apparent. Most glasses do not transmit well below about 300 nm (borosilicate for example has only a 50 per cent transmission at 300 nm) and in the far UV the absorption of air is appreciable over a few metres path length. So some constraints are obvious to start with. We need UV-transmitting glasses, fused silica is good down to about 170 nm, or sapphire to 140 nm. For detection down to about 110 nm a LiF window can be used. Beyond this the window is removed entirely.

One of the simplest devices used to detect UV radiation is the vacuum photodiode, which is a photoemissive device operating under vacuum conditions. This device must not be confused with the solid-state photodiodes discussed below.

Fundamentally, a photoemissive device is one that will emit a photoelectron from its surface after absorption of a suitable photon. This can be described by consideration of the photoelectric effect (discussed below) and by taking account of the work function of the material. The direction of the emitted photoelectron does not matter greatly, because an electric field can capture and control it.

A vacuum photodiode is illustrated in Figure 6.11. Ultraviolet radiation is incident on the photocathode, which causes photoelectrons to be emitted from the surface. An electric field enables the photoelectron to be captured and guided to the anode. This causes a current to flow between anode and cathode that depends on the intensity of the incident radiation. The anode illustrated in Figure 6.11 is long and thin so as not to shadow the photo cathode from incident radiation.

We have to select the appropriate material for the photocathode such that the incoming UV photon is sufficiently energetic for the generated electron to be liberated from the surface, that is to overcome the work function of the material. Many metals will not respond to the red or infrared regions of the spectrum, while some will not respond to visible radiation at all. Thus the photocathode and the associated window materials have to be selected for complementary characteristics.

The combination of a CsTe photoemitter with an LiF window enables a response at wavelengths as low as 110 nm.

One key parameter in the selection of a suitable photocathode material is its quantum efficiency, which is defined as the ratio of the mean number of photoelectrons

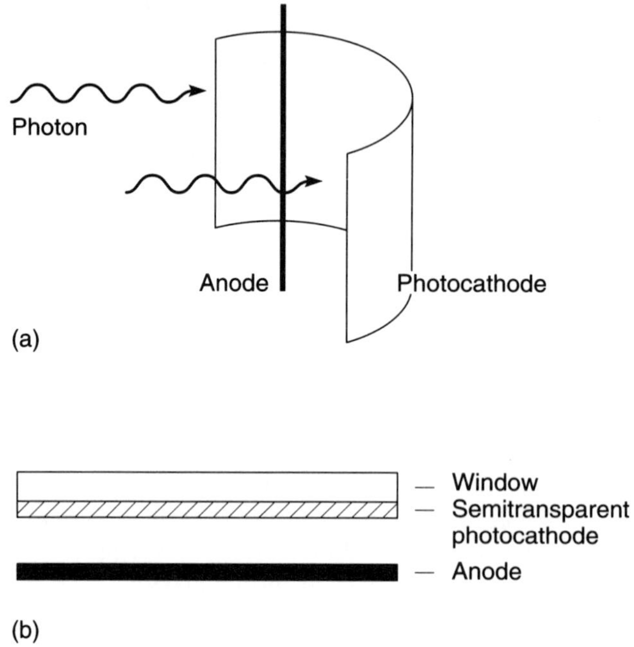

Figure 6.11 Vacuum photodiodes: (a) cylindical type, (b) planar type

released to the mean number of photons required to produce them. Most suitable materials have quantum efficiencies of less than 25 per cent.

The responsivity of a vacuum photodiode is low and will be influenced by any gas molecules between the anode and cathode that might absorb the photoelectron; hence the diode is under vacuum.

Table 6.7 gives an indication of photocathode materials available, responsivity and the wavelength range obtained with common window materials.

Table 6.7 Vacuum photodiodes: materials and characteristics

Material	*Spectral range* *nm*	*Sensitivity* *A/W*
CsTe	185 to 320	0.015
CsSb	185 to 650	0.06
NaKCsSb	185 to 850	0.04
GaAsP	200 to 800	0.18 (solid state)

The speed of response really only depends on the time of flight of the photoelectron, which under high field conditions can be very short and for planar devices can be of the order 170 ps. A planar device has the anode and photocathode close together and hence the time of flight is much reduced and the speed of response increased. A high field strength, of the order 1.5 kV across a 1.5-mm gap, further increases the speed of response.

Table 6.7 gives an indication of the sensitivity of detectors of the vacuum photodiode type. If more sensitivity is required then a photomultiplier must be considered. Photomultiplier tubes come in a variety of different designs, and use either secondary emission dynode or a microchannel plate construction. They are characterized by a very high gain, typically of the order 10^5 to 10^8, and hence find application where incident radiation levels are low.

The photomultiplier consists of a photocathode followed by an electron multiplier consisting of a series of secondary electron emission electrodes (dynodes) and a final collector which is the anode (see Figure 6.12).

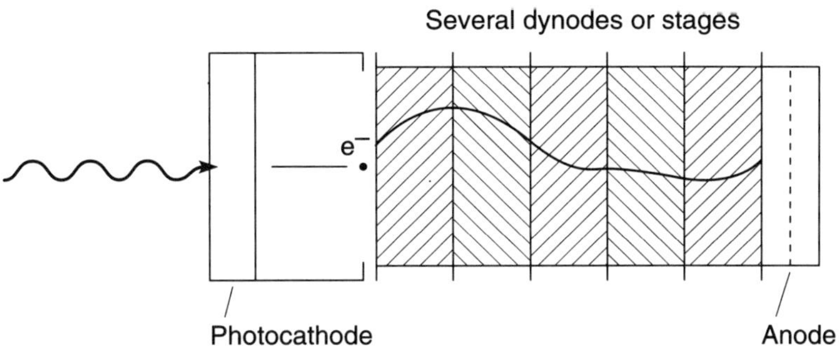

Figure 6.12 The photomultiplier (Venetian blind type)

There are two main types, the focused and Venetian blind constructions, each with relative advantages and disadvantages as shown in Table 6.8. Again the criteria of a UV transmitting envelope or window together with vacuum apply.

Table 6.8 Properties of photomultipliers

Focused	*Venetian blind*
Well-defined electron path	Poorly defined path Rugged
Fast rise time (1.5 ns)	Slow rise time (20 ns)
Fast transit time (5 ns)	Slow transit time (100 ns)
Narrow pulse widths	Wide pulse widths
Short life	Good stability

The operation of the photomultiplier depends on the acceleration of the photoelectron from the cathode to the first dynode, secondary electrons from the first dynode to the second, from second to third and so on until from the last dynode to the anode. The acceleration is achieved by establishing a potential difference (electric field) between each stage and the next. For each incident photoelectron a larger number of secondary photoelectrons are emitted and hence gain builds stage by stage.

A typical photomultiplier used for UV-visible measurements could have a specification as follows:

Spectral range	185 nm to 650 nm
Cathode material	SbCs
Window material	UV glass
Cathode sensitivity	48 mA/W
Anode sensitivity	1.2×10^6 mA/W
Rise time	about 2 s
Anode-to-cathode volts	1000 V d.c.

However, many types of photomultiplier are available with a range of different spectral sensitivities and rise times, so it pays to shop around. Photocathode materials and windows are similar for vacuum photodiodes and photomultipliers.

An operating photomultiplier tube should be protected from bright lights, even room lighting can cause damage.

Visible

By definition, the only true detector of visible light is the human eye, and, as human eye performance varies from person to person, this is not ultimately reliable. To overcome this to some extent many human eye spectral responses can be measured and averaged and a mean response produced. This photopic response has been discussed in detail above; in order to obtain a fair approximation to it, a detector material such as silicon has its spectral response modified by a filter.

What materials are suitable for forming the basis of a visible-light detector? There is only one material in common use, the ubiquitous silicon. A few years ago selenium would have figured as a major player but costs, performance and availability have sidelined this particular material.

Silicon has a response from the near ultraviolet (UV) through to the near infrared (IR). A spectral response curve is presented at Figure 6.13. Now, concentrating on silicon detectors, how do they work and what are their characteristics? Read on!

The photoelectric effect

Modelling or description of the photoelectric effect relies upon the particle theory of light. A simple statement of this theory is that light energy is carried by the photon, and that this fixed parcel of energy has a magnitude of:

$$E = h\nu$$

where E = energy
h = Planck's constant
ν = frequency of light.

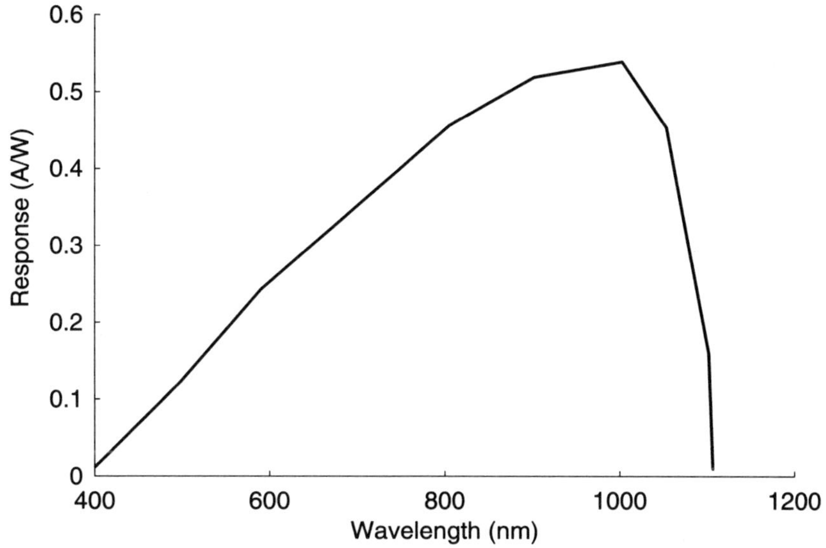

Figure 6.13 Silicon spectral response

When a suitable photon is incident on and absorbed by a material exhibiting the photoelectric effect, an electron is liberated and made available to conduct an electrical current. The electron is liberated only if the photon is of the correct energy. Long-wavelength, and hence low-frequency, photons do not carry sufficient energy to liberate electrons from silicon, and hence from a wavelength higher than about 1.1 μm its responsivity is zero .

Increasing the intensity of the incident light increases the number of photons falling on the surface and increases the number of electrons liberated. Note that, with an increase in intensity, it is only the number of photons that is increased, not the energy they carry, the latter being a fixed amount.

Very simply, a photon absorbed by a material gives its energy to an electron, which is then liberated from its bound state. This electron is then available for conduction through the material, or, if liberated from the material itself (photoemission), is available for transport to an anode.

The p-n junction

The p-n junction is discussed as a source of photons in Chapter 8 on light sources. Here we look at how a p-n junction can be used as a detector of incident photons. The p-n junction is the transition boundary between p-type and n-type materials in a semiconductor; it can be used as a simple photodiode.

The doping of a semiconductor with impurities establishes p- and n-type regions where, respectively, holes and electrons are available for conduction purposes. Between these two regions is the junction (itself a region not a plane), where the resultant electric potential acts as a barrier to current flow. Simply, incident photons result in the lowering of the potential barrier so that current flow can take place; the more numerous are the photons, the more current flows.

Reverse biasing the junction raises the potential barrier, while forward biasing lowers it. Only a few volts of bias are required to lift the barrier so that no free carriers can cross the gap. Statistically, and depending on temperature, some electrons are generated in the p-type material and some holes in the n-type material and a small reverse current will exist. This is the photodiode dark current and is highly temperature-dependent.

Forward biasing (where a negative supply is connected to the n-type side of the junction) allows high current flow and the rapid recombination of electron–hole pairs with the attendant release of energy as photons. This is an LED or semiconductor laser.

At an unbiased junction an incident photon liberates an electron from the valence band to pass down the potential barrier and recombine. This allows a photon-density-dependent current to flow – a photodetector.

Reverse biasing the junction (where a negative supply is connected to the p-type side of the junction) has the effect of reducing the junction capacitance and of allowing a much faster speed of response. Greater powers, and hence signal strengths, are allowed also, but these are dependent on the magnitude of the reverse bias. As with the unbiased junction, holes and electrons flow down the potential barrier for recombination and hence it is reverse current that flows. However, owing to the reverse biasing of the circuit, polarity is reversed between the unbiased and reverse-biased cases.

The voltage characteristic of a p-n junction is illustrated in Figure 6.14.

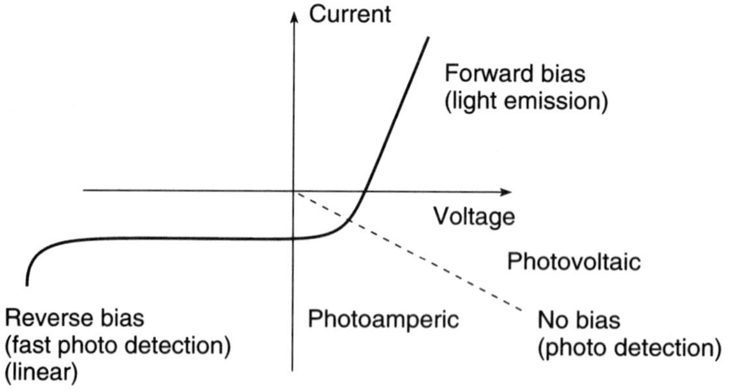

Figure 6.14 Voltage characteristic of the p-n junction

The PIN photodiode

The PIN photodiode is much the most popular photodetector available for general purpose detection. The 'I' in PIN stands for 'intrinsic' and represents the addition of an intrinsic (not doped) layer between the p- and n-type regions of the semiconductor junction.

The addition of the intrinsic layer reduces the capacitance of the depletion region still further than that of the reverse-biased p-n junction. This, together with reverse biasing, increases the speed of response and improves linearity.

Rise times of the order of a nanosecond are possible with small area devices. The device is likely to be manufactured with a thin, virtually transparent p-type silicon, which is protected by an antireflection coating. Photons will penetrate this surface such that absorption occurs in the extended depletion region.

The salient characteristics and features of the PIN photodiode that a specifier should be aware of are given below. The reader should be aware that there are as many ways of specifying a photodiode as there are manufacturers. The published characteristics can take a little unravelling!

Responsivity

Detectors for the visible respond outside of the visible spectrum and hence it is usual to quote this variable in A/W. The figure tells us how many amps of current will flow from the detector for the Watts of radiant power falling on it.

Due to the spectral dependence of this parameter, it is usual to have it quoted at one wavelength or to supply a graph of the response. Typically, near the peak of response, the responsivity should be of the order of 0.5 A/W (silicon).

Spectral range

The useful spectral range of silicon extends from about 0.4 μm through to about 1.1 μm. This can be modified by filters deposited on top of the receiving surface of the diode. We have already mentioned the photopic response filter that can be formed on the detector, but virtually any filter response can be generated.

The protective window over the detector itself can be made from a variety of different materials, each of which will affect spectral response by virtue of its spectral transmission characteristic. For example, to enhance UV response a quartz or other UV transmitting window must be used.

Noise limits

Usually the dominant noise mechanism is that of Shot noise (dark leakage current) closely followed by that of Johnson noise from the resistive elements.

When discussing noise from a detector, most designers are really interested in the minimum power that the detector can reliably recognize. What is required then is an estimate of the minimum incident power that will equate to the noise of the detector. This is called the noise-equivalent power (NEP), and can be expressed as

$$\text{NEP} = \frac{\text{noise current (A)}}{\text{photodiode sensitivity (A/W)}}$$

Some manufacturers will quote detectivity (D) and not NEP: $D = 1/\text{NEP}$. The NEP should be quoted at a specific wavelength (photodiode sensitivity varies with wavelength) and for the specific frequency range over which the noise was measured. The unit of measurement of NEP is the watt (W) and that of D the reciprocal watt (W^{-1}).

The noise-equivalent power depends on the bandwidth of the measuring system and the surface area of the detector. In order to account for these variables, and to enable a meaningful comparison between detector types, we need to consider D^* ('D star').

$$D^* = A^{1/2}B^{1/2}/\mathrm{NEP}$$

where A and B refer to the detector area and bandwidth respectively. The unit of measurement of D^* is quite indigestible, and is given in

$$D^* = \mathrm{cm\ Hz^{1/2}\ W^{-1}}$$

You will often see D^* quoted for a particular source temperature, for example 'D^* (500)'.

Speed

A square wave of light applied to a PIN photodiode will not be transformed to an electrical square wave, unless you have a fast detector and fast-processing electronics.

The application of a reverse bias to the PIN photodiode increases the size of the depletion region and creates a field across the diode that accelerates the charge carriers across the region for collection. Thus a reverse bias will increase the speed of response of the diode. There are two components to consider; those carriers created in the depletion region are collected quickly as they are swept along by the field. However, those created in the bulk of the device, the non-depleted region, have to drift toward collection. Thus there are a fast and a slow characteristic to the response. See Figure 6.15.

If the reverse bias is increased then the depletion region increases in depth, with the advantage that more carriers are created in the depletion region and the field strength

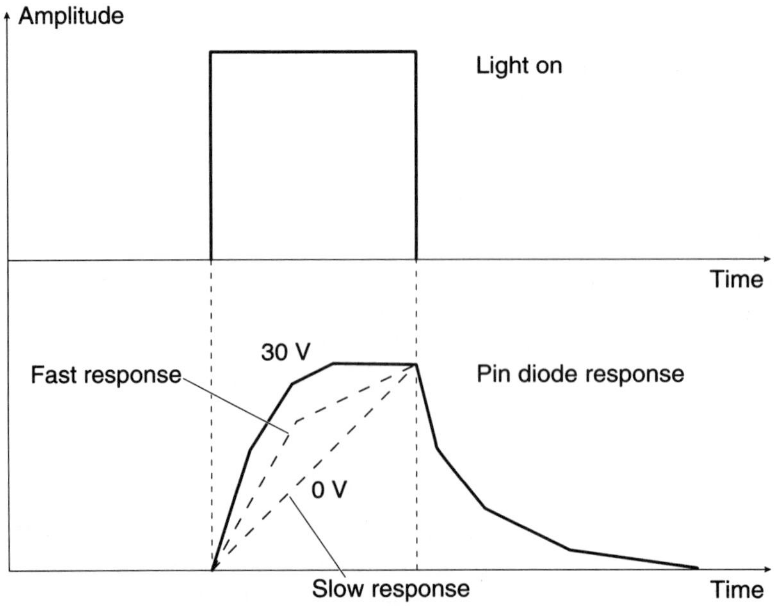

Figure 6.15 Speed of response

is higher, enabling still faster collection. There is a limit to how large a bias can be applied. Ultimately the depletion region encompasses nearly all of the diode and physically cannot be made any larger. The only further gain is through increased field strength and hence the acceleration of carriers. Ultimately, with too high a field strength the device will break down.

It is worth noting that the longer is the wavelength of the incident light the greater will be the penetration depth. Light of a wavelength about 450 nm will penetrate silicon to a depth of about 1 μm, that of 750 nm to about 10 μm. Beyond about 1.2 μm silicon is essentially transparent, and makes a good optical material for infrared lenses.

Temperature

Temperature affects the operation of a PIN photodiode in two ways. First, there is a exponential increase in thermally generated electron pairs that increases leakage current. Leakage current approximately doubles for about every 10 °C rise in junction temperature. Second, the opacity of silicon increases with temperature, enabling more long-wavelength photons to be captured and hence increasing longwave efficiency. Conversely, UV collection efficiency falls.

The avalanche photodiode

If the field strength, due to the applied reverse bias, across a PIN diode is high enough and the path of a free carrier long enough then there is a good chance that the free carrier will become energetic enough to 'knock out' or liberate another carrier. These two become four, the four become eight, and so on. This is the principle of the avalanche photodiode.

An avalanche photodiode (APD) requires high bias of the order 300 V to 500 V, which is not always convenient. The temperature-dependent bias level needs careful and accurate control. These two factors make the associated electronic circuitry more complicated than for the PIN diode.

Gains of the order of 50 to 300 are achievable with only a small increase in noise, and hence the signal-to-noise ratio is improved.

Infrared

For our purposes here we take the infrared to mean those regions exploited by fibre optic communications and the two atmospheric windows that allow infrared transmission down to sea level. Fibre optic transmission windows for communication purposes can loosely be centred on 0.9 μm, 1.3 μm and 1.5 μm. The atmospheric transmission windows are from approximately 3 to 5 μm and from 8 to 14 μm. These are the bands most useful for terrestrial thermal imaging.

The most common detector used with fibre optics for communications is the PIN photodiode already discussed above. Broadly, silicon is the base material for the shorter 0.9 μm applications, whereas GaAs (gallium arsenide) is used for the two longer-wavelength regions.

A common detector of 'thermal' infrared radiation is the pyroelectric detector. The detector is a current source with output proportional to rate of change of temperature. The detector can be operated at room temperature (300 K), and is usable over the wavelength range 0.6 to 35 μm with a D^* (500) of about 8×10^8 and a time constant of about 0.1 to 100 ms. Compare this with a silicon APD, which will also operate at room temperature over a wavelength range of 0.6 to 1.1 μm with a D^* (500) of $> 2 \times 10^7$ and a time constant of the order of 10^{-5} μs. So the pyroelectric is relatively slow and needs a.c. coupling, but is not so bad in terms of responsivity.

The so-called lead salt detectors, such as PbSe (lead selenide) and PbS (lead sulphide), are similar in some respects to the pyroelectric type, but have superior detectivity over a limited wavelength band.

A cooled thermal detector generally gives a superior response in terms of detectivity. Cadmium mercury telluride, CdHgTe (CMT or occasionally MCT!) is one of the most popular and, depending on the cooling arrangements, is usually operated at either 195 K or 77 K. Cooling is often achieved through a Dewar flask filled with liquid nitrogen.

Table 6.9 compares some of the detectors mentioned above.

Table 6.9 Properties of photodetectors

Detector	*Temperature* K	*Time constant* ms	*Usable wavelength* μm	D^* *(500)* cm $Hz^{1/2}$ W^{-1}
Pyroelectric	300	0.1 to 100	0.6 to 35	$\leqslant 8 \times 10^8$
Silicon (APD)	300	approx. 10^{-5}	0.6 to 1.1	$>2 \times 10^7$
PbSe	300	>100	1.1 to 3.8	1.5×10^9
CdHgTe	195	approx. 2	1 to 5.5	4.0×10^{10}
CdHgTe	77	0.1 to 2	8 to 14	2.5×10^{10}

CCD camera

Charge-coupled devices (CCDs) are now very commonly used for CCTV and even some broadcast TV situations. In the simplest terms, an array of photo receivers store charge created by the photoelectric effect. At the end of an integration period, usually 20 ms for a CCTV application, the charge is transferred sequentially to a shift register, where full lines are clocked out.

Figure 6.16 illustrates the operation of an interline transfer CCD. At the end of an integration period, charge is transferred from the integration region to a dark transfer region. This frees the integration region, leaving it ready to start charge collection for the next readout. The charge that has transferred from integration to transfer is now clocked, row by row, up towards the shift register. As each line appears at the shift register it is clocked out for processing as a video line. The transfer region is dark so as to prevent corruption of the charge packet through further generation via the photoelectric effect.

Use of three CCDs with a colour-splitting prism will give a high-resolution colour TV camera. A single CCD but with a different colour filter on each picture element (pixel) results in a colour TV camera with a lower resolution.

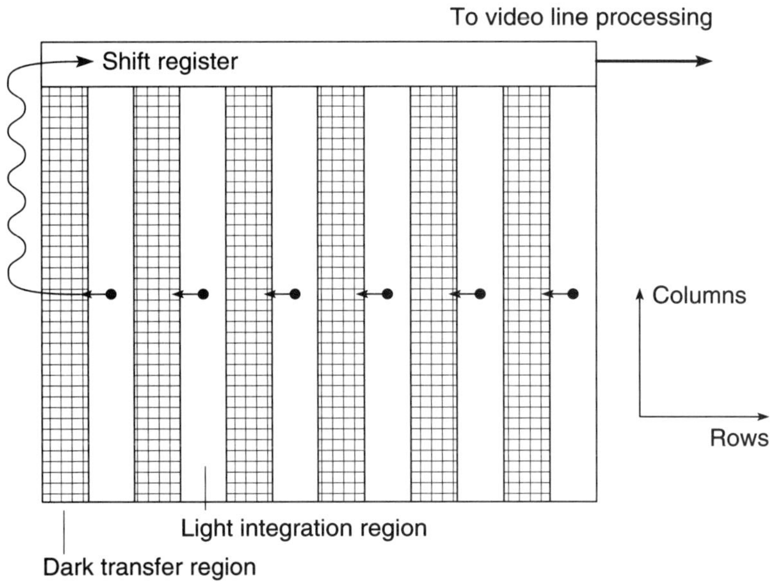

Figure 6.16 CCD camera chip. At the end of an integration period charge is transferred from the light integration region to the shielded transfer region. During the next integration period the charge stored in the transfer region is clocked up to the shift register, line by line for output for video line processing

Thinning and cooling of the CCD chip gives a TV camera that exhibits very low noise over very long exposure times, of the order of hours. This finds application for long exposures for astronomy purposes.

Reverse clocking the chip during exposure 'dumps' excess charge. Thus if a chip is reverse-clocked for, say, 19 out of 20 ms then a 1-ms exposure TV camera has been achieved, which is useful for freezing action.

The fixed and very stable geometry of the CCD camera makes it an ideal choice for industrial inspection where lack of distortion, reliability and ruggedness are paramount. In reply to the question, 'Did the CCD camera revolutionize automatic visual inspection?', we think it did!

Detector systems and good housekeeping

Some of the most important features of detector systems are with respect to the control of stray light. Stray light can be the cause of so many problems, fluctuating signals, distorted distributions, low signal-to-noise, or whatever. There are some measures that can be taken to avoid these pitfalls and a few ideas are presented here.

The 'good housekeeping' in the heading above refers to techniques that are essentially common sense, but occasionally need pointing out. One such piece of 'good housekeeping' relates to the use of lasers in measurement systems. Lasers offer an intense, controllable and stable source of light, but signal detection can often be confused by contributions from ambient light. A narrow-band filter centred on the laser wavelength will serve to considerably reduce this contribution and makes for a more robust detection system.

To keep the advice simple and concise we offer the following list of rules, which we are sure you will be able to add to.

Principles of good photometric and radiometric housekeeping

1. Use a calibrated instrument! There is no point whatever in attempting to make an absolute photometric measurement with an instrument that has a dubious calibration. Comparative or ratiometric measurements can be made, but don't assign real numbers.
2. Control stray light in the measurement laboratory. Blackout the windows to prevent daylight influencing the measurement. Look for stray reflections from walls, ceiling, floor and any part of the measurement apparatus. Incidentally, do record your measurement arrangements, so that if you have to repeat them you can reproduce your original setup exactly.
3. Control stray light within the measuring instrument itself. For example, if you are building your own radiometer use baffles, often concentric rings, to control the field of view (see Figure 6.17). Make sure that the baffles are painted matt black. Two other options to assist in the control of stray light within an instrument are fine screw cutting the inside of the bore or lining with black felt. The black felt is particularly good for control of light at grazing incidence.

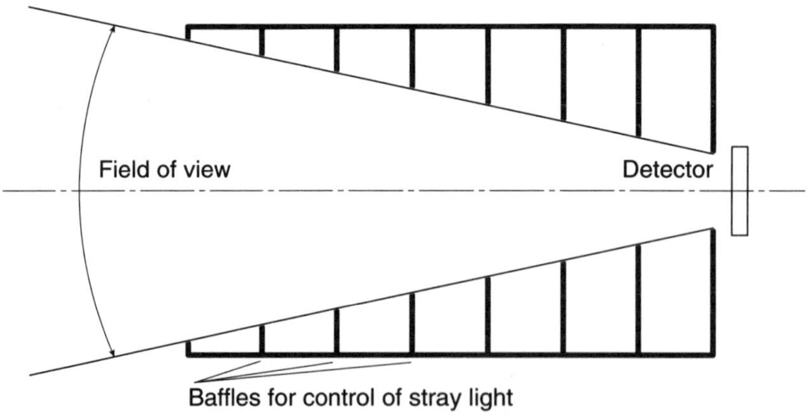

Figure 6.17 Baffles for the control of stray light

4. Use of optics can considerably increase the flux presented to a detector, in the same way as a large-aperture telescope presents a brighter image to the eye (see Figure 6.18). However, as illustrated in Figure 6.18, the use of a lens will significantly reduce the field of view of the radiometer. This is of course desirable for some applications, for example measurement of projection screen brightness at a point.

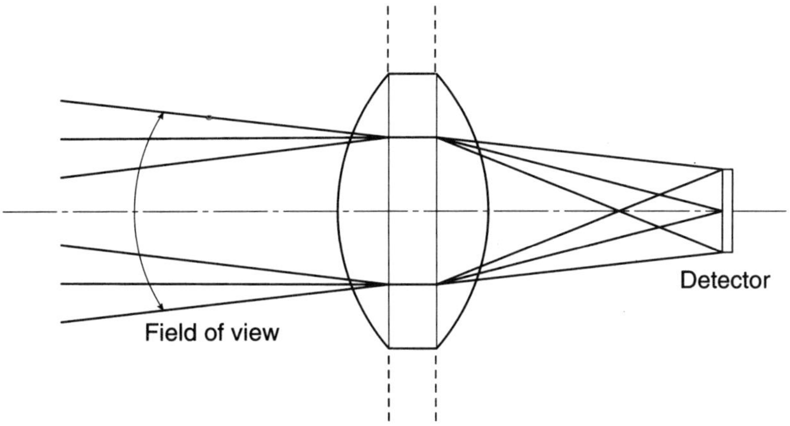

Figure 6.18 Lens to control field of view

5. The field of view of a lensed radiometer can be increased through the use of a field lens. Indeed, if the radiometer is to interface to another optical instrument, such as a monochromator, then the apertures of the instruments can be matched through the use of a field lens. See Figure 6.19, which illustrates the basic principle.
6. If lenses are used in your detection system then unless they are coated you will lose approximately 4 per cent of the light per lens surface (for the visible region of the spectrum). Thus if an objective lens and field lens are employed only 85 per cent of the incident light will be transmitted. This can be largely avoided through use of appropriate coatings.

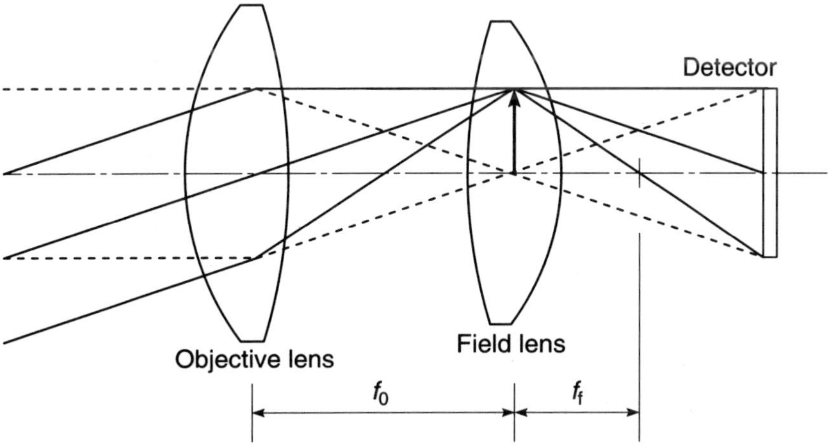

Figure 6.19 Increased field of view through use of a field lens

TABLE 6.10 CIE colour matching functions

Wavelength nm	$x(\lambda)$	$y(\lambda)$	$z(\lambda)$	*Wavelength nm*	$x(\lambda)$	$y(\lambda)$	$z(\lambda)$
380	0.0014	0.0000	0.0065	585	0.9786	0.8163	0.0014
385	0.0022	0.0001	0.0105	590	1.0263	0.7570	0.0011
390	0.0042	0.0001	0.0201	595	1.0567	0.6949	0.0010
395	0.0076	0.0002	0.0362	600	1.0622	0.6310	0.0008
400	0.0143	0.0004	0.0679	605	1.0456	0.5668	0.0006
405	0.0232	0.0006	0.1102	610	1.0026	0.5030	0.0003
410	0.0435	0.0012	0.2074	615	0.9384	0.4412	0.0002
415	0.0776	0.0022	0.3713	620	0.8544	0.3810	0.0002
420	0.1344	0.0040	0.6456	625	0.7514	0.3210	0.0001
425	0.2148	0.0073	1.0391	630	0.6424	0.2650	0.0000
430	0.2839	0.0116	1.3891	635	0.5419	0.2170	0.0000
435	0.3285	0.0168	1.6230	640	0.4479	0.1750	0.0000
440	0.3483	0.0230	1.7471	645	0.3608	0.1382	0.0000
445	0.3481	0.0298	1.7826	650	0.2835	0.1070	0.0000
450	0.3362	0.0380	1.7721	655	0.2187	0.0816	0.0000
455	0.3187	0.0480	1.7441	660	0.1649	0.0610	0.0000
460	0.2908	0.0600	1.6692	665	0.1212	0.0446	0.0000
465	0.2511	0.0739	1.5281	670	0.0874	0.0320	0.0000
470	0.1954	0.0910	1.2876	675	0.0636	0.0232	0.0000
475	0.1421	0.1126	1.0419	680	0.0468	0.0170	0.0000
480	0.0956	0.1390	0.8130	685	0.0329	0.0119	0.0000
485	0.0580	0.1693	0.6162	690	0.0227	0.0082	0.0000
490	0.0320	0.2080	0.4652	695	0.0158	0.0057	0.0000
495	0.0147	0.2586	0.3533	700	0.0114	0.0041	0.0000
500	0.0049	0.3230	0.2720	705	0.0081	0.0029	0.0000
505	0.0024	0.4073	0.2123	710	0.0058	0.0021	0.0000
510	0.0093	0.5030	0.1582	715	0.0041	0.0015	0.0000
515	0.0291	0.6082	0.1117	720	0.0029	0.0010	0.0000
520	0.0633	0.7100	0.0782	725	0.0020	0.0007	0.0000
525	0.1096	0.7932	0.0573	730	0.0014	0.0005	0.0000
530	0.1655	0.8620	0.0422	735	0.0010	0.0004	0.0000
535	0.2257	0.9149	0.0298	740	0.0007	0.0002	0.0000
540	0.2904	0.9540	0.0203	745	0.0005	0.0002	0.0000
545	0.3597	0.9803	0.0134	750	0.0003	0.0001	0.0000
550	0.4334	0.9950	0.0087	755	0.0002	0.0001	0.0000
555	0.5121	1.0000	0.0057	760	0.0002	0.0001	0.0000
560	0.5945	0.9950	0.0039	765	0.0001	0.0000	0.0000
565	0.6784	0.9786	0.0027	770	0.0001	0.0000	0.0000
570	0.7621	0.9520	0.0021	775	0.0001	0.0000	0.0000
575	0.8425	0.9154	0.0018	780	0.0000	0.0000	0.0000
580	0.9163	0.8700	0.0017				

TABLE 6.11 Conversion photometric and radiometric quantities

Measurement	*Photometric*	*Radiometric*	*Symbol*
Energy	Luminous energy lumen second talbot lm s	Radiant energy watt second joule W s	Q
Energy density	Luminous density $\mathrm{lm\,s\,m^{-3}}$	Radiant density joule	w
Power	Luminous flux lumen lm	Radiant power watt W	Φ
Flux density	Illuminance luminous exitance lux or lx $\mathrm{lm/m^2}$	Irradiance radiant exitance $\mathrm{W\,m^{-2}}$	E M
Intensity	Luminous intensity candela lumen per steradian $\mathrm{lm\,sr^{-1}}$ cd	Radiant intensity watt per steradian $\mathrm{W\,sr^{-1}}$	I
Brightness	Luminance nt $\mathrm{cd/m^2}$	Radiance $\mathrm{W\,m^{-2}\,sr}$	L
Luminous efficacy	lumen per watt $\mathrm{lm\,W^{-1}}$		K
Spectral luminous Efficacy	lumen per watt $\mathrm{lm\,W^{-1}}$		$K(\lambda)$
Luminous efficiency	$K(\lambda)/K(\lambda_{max})$		$V(\lambda)$

CHAPTER 7

Fibre optics

Introduction

This chapter is simply intended to give an overview of the topic of optical fibres and their uses. It includes a description of the various types of fibre and the connectors used to join or terminate fibres to the outside world. Some practical discussion of fibre preparation and the means of getting light into the fibre is also presented. Finally some of the more common applications of fibres are described.

Fundamentals

In order to use light as an effective means of communication it must be pulsed to encode the information. Each pulse has to be transmitted over considerable distances without significant loss or distortion. This is achieved routinely in the laboratory over more than 100 km for a single repeaterless run, and it can only be done by using glass of a very high purity.

The highest grades of fibres, developed specifically for communications purposes, exhibit losses of only about 0.6 dB/km and laboratory developments show that 0.2 dB/km is possible, this is close to the reported theoretical limit.

To determine the optical power in decibels, at a given distance, for a fibre of known attenuation we use the formula

$$\text{attenuation} = 10 \log (\text{input power}/\text{output power})$$

For example: over what distance could we transmit a pulse so that we received 10 per cent of the input power using a fibre with an attenuation of 0.2 dB/km? The answer is given by

$$0.2 = 10 \log (P1/P2)$$

$$P2 = P1/10^{0.02}$$

$$P2 = 0.95\,P1 \text{ after } 1\,\text{km}$$

If the transmission is to be over x kilometres then

$$0.95^x = 0.1 \qquad \text{(for 10 per cent signal strength)}$$

$$x = \log (0.1)/\log (0.95)$$

$$x = 45\ \text{km (approximately)}$$

In reality, most fibre optic runs for land use will be shorter than this, because of the practicalities involved with actually laying the cable. However, for underwater applications, such as transatlantic cables, the very lowest attenuation will help to reduce the number of repeater stations required.

Nevertheless, with an optical fibre we are not looking through a 50-km slab of glass with a direct path for the light from end to end; the fibre is a waveguide, not a 50-km thick glass window. How does the light manage to get from end to end and go round all of the bends in the optical fibre?

To understand how this is achieved we need to recall the phenomeon of total internal reflection (TIR), first introduced in Chapter 3. TIR occurs at a refractive index boundary when the incident ray is travelling towards the lower-index material and the angle of incidence is greater than the critical angle. For these rays the refractive index boundary behaves like a perfect mirror. Remember that the angle of incidence is measured from the surface normal, so for a glass-air interface the angle beyond which TIR occurs is about 42 °.

Provided that ray angles are shallow enough at the index boundary, they will propagate without reflection loss along the length of the fibre. Figure 7.1 shows the passage of a meridional ray through a short fibre optic of uniform refractive index and with total internal reflection occurring at the n_1 to n_2 boundary. The geometric ray path is easy to analyse with the formula

$$\text{path length} = S/\cos U' \tag{7.1}$$

The number of reflections undergone by this ray in traversing the length of the guide is given by

$$\text{number of reflections} = S \tan U'/d \tag{7.2}$$

where d is the diameter of the guide or fibre. As stated above, these simple relationships are true for meridional rays only. However, both relationships serve to introduce some important concepts.

The path length depends on the refracted angle U', and thus on the angle of incidence and the refractive indices of the two parts. Glass is a dispersive medium, which means that the refractive index varies with wavelength. So the path length travelled by a ray will depend not only on its launch angle but also on its wavelength. This is referred to as dispersion, and is a cause of pulse broadening, where light of different wavelengths take different times to travel along a fibre optic.

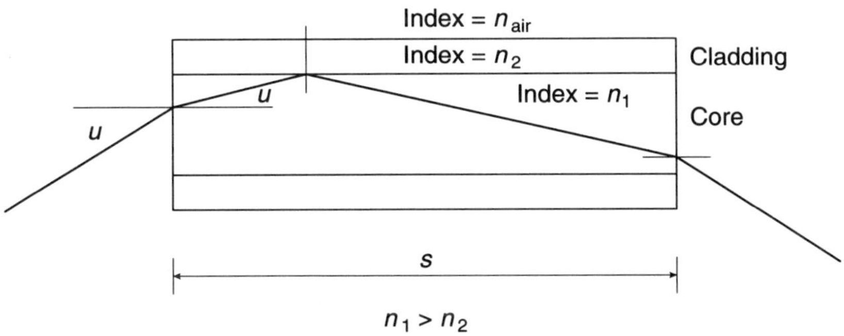

Figure 7.1 Simple fibre optic

Pulse broadening sets a fundamental limit to the distance over which two pulses can travel before they overlap to such an extent that they are indistinguishable from each other. The exact distance would of course depend on the degree of monochromaticity of the source, the distance between pulses and the pulse length.

For an optical fibre the critical angle is dependent on the ratio between the respective refractive indices of the cladding (n_2) and the core (n_1), and is given by

$$\text{critical angle} = \sin^{-1} n_2/n_1 \tag{7.3}$$

Given that glass is a dispersive medium, and that the critical angle depends on refractive index, it can be seen that certain wavelength and propagation angle combinations are less likely to be guided by the fibre than others. Very crudely, these can be referred to as leaky modes, and a fibre will strip these modes from propagation over the first few metres.

For example, the glass Schott BK7 has refractive indices of 1.51 and 1.53 in the red and blue respectively. At a glass–air boundary the magnitude of the critical angle in each case will be

red	$\sin^{-1} 1.00/1.51 = 41.47°$
blue	$\sin^{-1} 1.00/1.53 = 40.81°$

Thus a distinct limit is set to the range of input angles that a fibre will be able to support and guide. This limit is usually referred to as the numerical aperture (NA) of the fibre, and is given by $n_{air} \sin^{-1} U$. This is further characterized by

$$\text{NA} = (n_1^2 - n_2^2)^{1/2} \tag{7.4}$$

This argument is all very well until the fibre bends. At a bend in the fibre the angles of incidence at the refractive index boundary can decrease and hence frustrate the TIR. Thus rays can be lost at bends. This loss can occur for very small distortions of the fibre (micro-bending losses) up to fairly large radii. A manufacturer will usually specify a minimum bend radius below which excessive losses will occur.

Lastly in this section we should take a quick look at the transmission spectrum of a fairly standard grade of communications fibre (see Figure 7.2). This figure illustrates

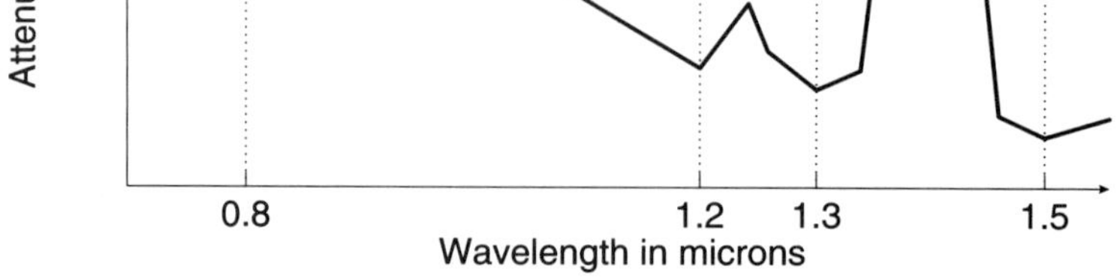

Figure 7.2 Generalized fibre transmission spectrum

the three regions of interest where acceptable transmission is achieved. For short-haul and relatively high attenuation the 'red' end of the spectrum can be used, whereas for longer distance communications use the 1.32 μm and 1.55 μm optical 'windows' are more appropriate.

You will of course need sources and detectors to exploit these windows. The ubiquitous silicon for the shorter wavelengths and GaAs for the longer are appropriate materials.

Incidentally, not shown explicitly in Figure 7.2 is the high blue attenuation exhibited by fibre optics (and glasses in general). Fibre optics are good at stripping out the blue content of a broadband source, and for illumination purposes this can be important.

Fibre types

There are five main types of fibre optic, each with their own niche market. They will be considered in turn below and comprise step-index, multimode, single-mode, plastic-clad silica and plastic-on-plastic types. Refer first to Figure 7.3 for the basic terms used in the description of the different types of fibre.

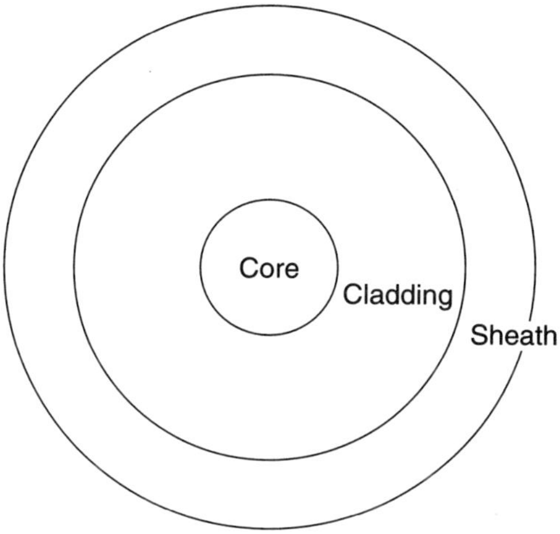

Figure 7.3 Basic fibre terms

Step index

In essence the descriptions above refer to a step-index fibre. This is simply a thin rod of glass, plastic, quartz, etc., clad with a second material or simply exposed to air. Some of its characteristics have already been described above, but some further points need to be discussed here.

We have already introduced the concept of different paths through a step-index fibre. These paths depend on a whole variety of different variables, which include launch angles and wavelength. In practice, a step-index fibre can support many such modes – hundreds, even – and this leads to the phenomenon of modal dispersion.

With reference to Figure 7.4, it is clear that ray B is going to take more time to traverse the fibre than ray A. Thus, from a single source delivering a short pulse of light there will be a range, or distribution, of arrival times at the far end of the fibre. This distribution leads to a broadening of the pulse, and in extreme cases, as illustrated at Figure 7.4 by the four discrete pulses, will cause a smearing out of the signal. These effects severely limit the application of step-index fibres to all but the simplest communications applications.

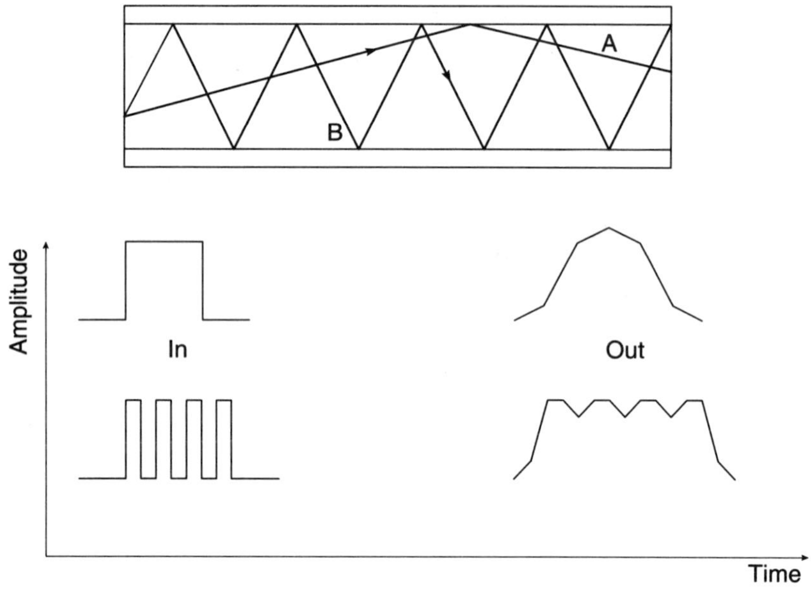

Figure 7.4 Modal dispersion

The step-index fibre really excels, however, as an efficient means of transporting light from one place to another, and finds wide application in many scientific areas such as endoscopy, microscopy and others. The fibre core is generally quite thick with a thin cladding, with only a small difference between the indices of the core and cladding materials. This makes for a usefully large NA together with easy coupling and connecting.

With a step-index fibre exposed to air, that is, with no cladding material, in essence a long thin glass (or plastic) rod, it is essential to keep the surface clean and free from grease, because this will serve to frustrate the TIR.

Multimode

The multimode optical fibre (sometimes called a graded-index fibre) employs a refractive index profile as illustrated at Figure 7.5. Light rays bend toward the region of higher refractive index and travel faster in regions of low refractive index. These two phenomena are put to good application with the multimode fibre.

Each ray traversing the length of a multimode fibre follows a gently undulating

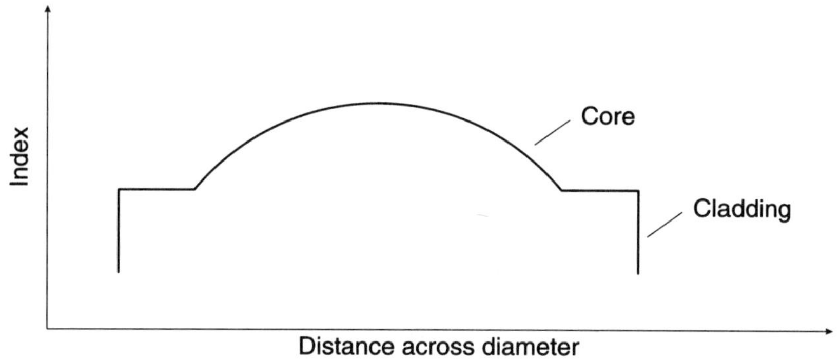

Figure 7.5 Multimode fibre refractive index profile

reciprocal course (see Figure 7.6). Those rays that reach the extreme edge of the core pass through material with a lower refractive index than do those traversing the centre region, and therefore travel a little faster, thus compensating for the increased physical path.

The refractive index profile, usually parabolic, effectively causes a lensing effect within the fibre, where all modes follow a sinusoidal path of the same pitch. In fact, rays originating from a single point will focus periodically along the length of the fibre. This is the principle of the GRIN rod lens. The refractive index profile can be defined as follows:

$$n(r) = n_{max}[1 - \text{const.}(r/2)^2] \qquad (7.5)$$

where r = radial distance from axis of fibre
n_{max} = axial value of refractive index
const. = a constant that depends on the variation of index and fibre diameter.

In this way the multitude of modes supported by the fibre can travel together with a virtually eliminated modal distortion.

The graded-index fibre is therefore capable of supporting high data rates over extended distances. The core and cladding diameters are not too small to work with,

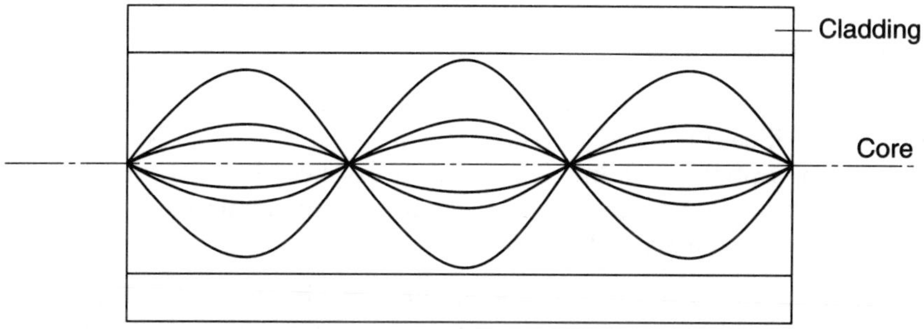

Figure 7.6 Ray paths in a multimode fibre

one standard being a 50/125 fibre, which means a core diameter of 50 μm and a cladding diameter of 125 μm. Coupling into and out of the fibre are both still relatively simple, as is jointing the fibre. The fibre typically has a relatively large NA, which enables good light gathering.

In terms of cost, the graded-index fibre is significantly more expensive than a step-index fibre.

Monomode

The monomode fibre, as the name implies, is capable of supporting only one propagation mode. This is achieved by reducing the diameter of the core until only one mode can be supported, and typically this diameter will be of the order 5 μm. The fibre is of the step-index sort.

The cladding of a monomode fibre has to be quite thick, because much of the energy propagated by the fibre will travel as an evanescent wave in the cladding. An evanescent wave is one that travels outside of the constraining refractive index boundary. There is by definition no modal dispersion, and hence this is the fibre used for long-distance communication purposes.

Monomode fibres are fragile, difficult to launch light into, with an NA of the order 0.1, and difficult to connect or join. With such a low NA, the rays travelling along the fibre are essentially collimated (see Figure 7.7). Monomode fibre is expensive.

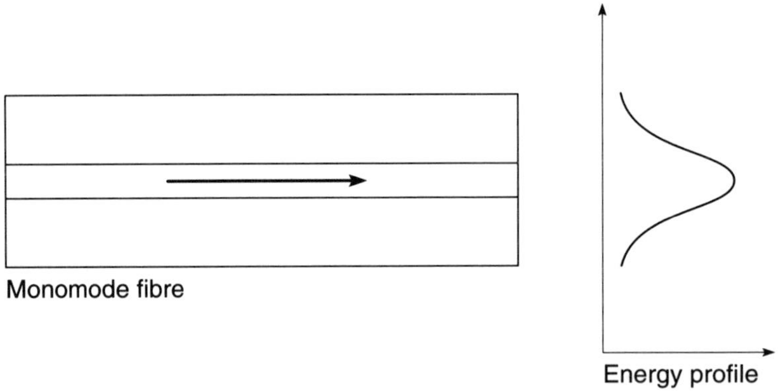

Figure 7.7 The monomode fibre

Plastic-clad silica

Plastic-clad silica (PCS) step-index optical fibres are mainly manufactured from synthetic fused silica or fused crystal, as core materials and silicone resins, or fluorocarbons, as optical cladding. They are in essence the same as the step-index fibre discussed above. However, they are cheap, readily available and have a host of different connectors, sources and detectors.

The performance of PCS fibres is adequate for simple, short-distance (500-m) communications, and hence they lend themselves very nicely to local area networks, etc.

Polymer fibre optic cable

This is perhaps the most readily available fibre optic system and is available from the major mail-order electrical wholesalers. The fibres are usually an acrylic optical cable sheathed with a fluorine that contains polymer. The cable is inexpensive and is ideal for demonstration of fibre optic systems, transmission link demonstrators and in some circumstances light pipes or illumination systems.

A typical specification would be

attenuation	150 dB/km at 600 nm
spectral range	400 to 700 nm
cable length	20 m
operating temperature	−30 to +80 °C
core diameter	1 mm
outside diameter	2.2 mm

The cable is easily cut to length with a scalpel or hot knife and the detector, source and connectors are usually packaged for simple snap-together operation.

Connectors

Connectors are an essential and important part of any fibre optic system. There is no point in pushing fibre manufacturing technology to the absolute limit in order to obtain the lowest loss figures if the connectors used in the system waste light.

The connector problem is of course different from one fibre type to another. Connecting two monomode cables with core diameters of 5 μm is a very different affair from connecting two fibres with core diameters of 1 mm. If the 5-μm core of a monomode fibre is displaced by 5 μm then you have a disaster on your hands, but if a 1-mm core is displaced by 5 μm, would you notice the difference in signal level?

The connector has to account for three possible misalignment scenarios (and maybe the combination of all three!):

1. Longitudinal separation of the fibre ends.
2. Orthogonal displacement of the axes of the fibres to be joined.
3. Angular misalignment of the fibre axes.

These misalignment scenarios are illustrated in Figure 7.8. It is the job of the connector to prevent these occurring. In the limit, end-to-end connections will have efficiency dictated by the Fresnel reflection losses at the glass–air–glass interfaces.

There are many other forms of connectors for straight end-to-end connection, for two-to-one, to star connectors, to wavelength-discriminating connectors. All of these have specific applications, and we cannot possibly present all of the detail required here. (The book *Opto-Electronics* by Chaimowicz gives a good overview and more detailed discussion.) However, we can present a simple overview as follows.

1. *Fusion splice.* This is a reliable, and now field-usable, technique for joining monomode and graded index fibres. Each end is prepared (cleaved, or cleaved and polished) and brought together under micrometer positioning control, until the

(a)

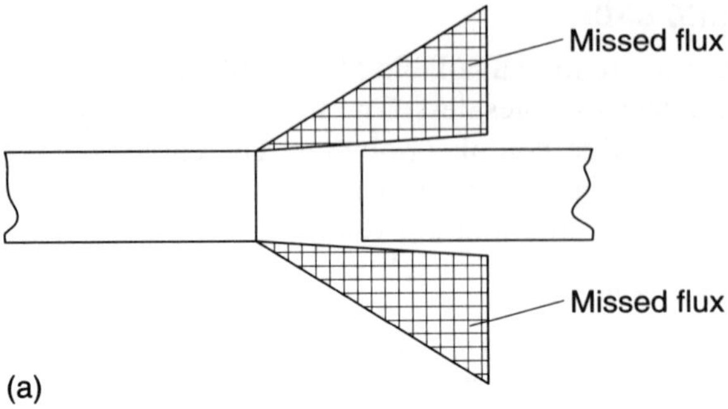

(b)

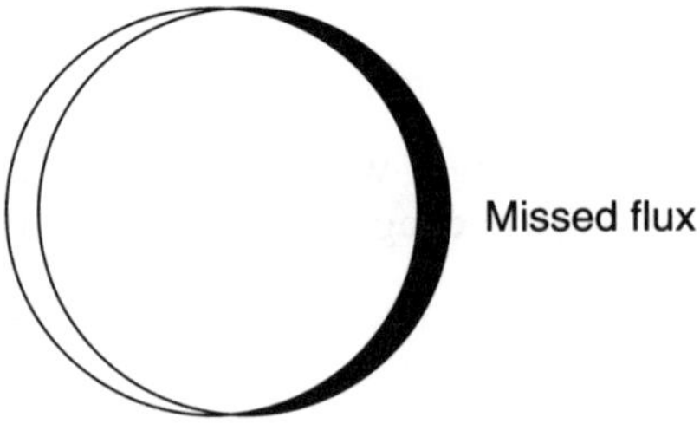

(c)

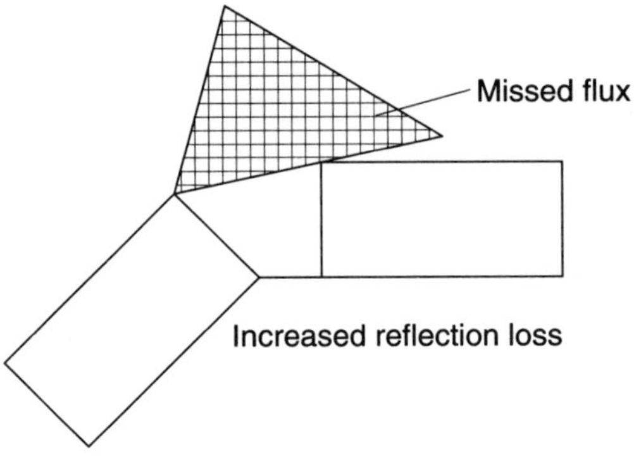

Figure 7.8 Fibre connector losses – three common types: (a) longitudinal separation, (b) orthogonal displacement, (c) angular misalignment

cores align in orthogonal directions. Observation is achieved with a microscope. Local heating is applied to the region, and the two ends literally fuse together, with surface tension effects helping to align the fibres.

This technique can result in splices with losses as low as 0.1 dB when used with 50/125-μm graded-index fibres. The microscope helps to position the fibre ends at the correct distance apart and with the correct relation to the heating electrodes.

This connection is not demountable and is not suitable for make-or-break-type connections.

2. *Fused biconical type.* This technique is particularly useful for the production of a 'Y' coupler, which consists of a single input fibre with two output fibres (see Figure 7.9). Two fibres are brought close together and their sheathing removed over a

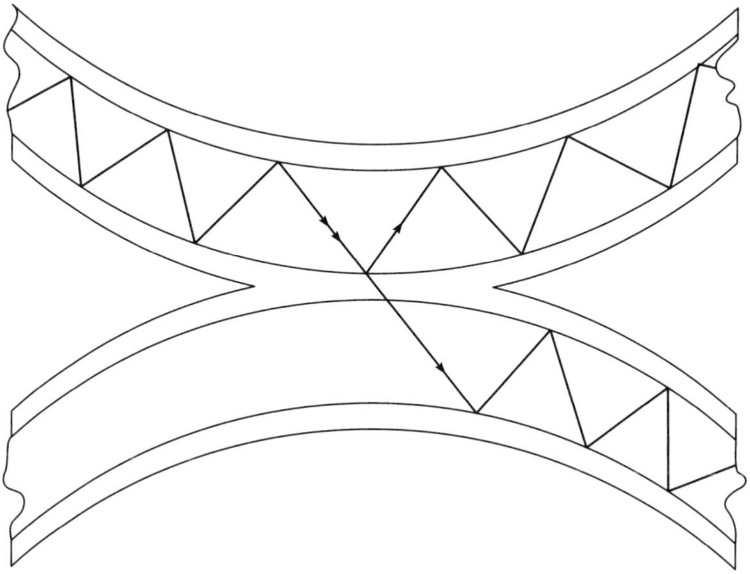

Figure 7.9 Fused coupler – biconical Y-type

short distance. The two 'sides' of the fibre are fused together so that some light can leak from the first fibre to the second. Get it right and you have achieved a 50:50 split.
This is not a demountable connection.

3. *Expanded-beam type.* This coupler takes advantage of collimating micro lenses that can create a collimated gap between two fibres. This eases some of the constraints with respect to longitudinal and orthogonal displacements.
The coupler can be formed from precision-aligned ball lenses, or the fibre end itself can be modified to act as a lens. With careful design and the use of alignment ferrules, this particular concept can be applied to reliable demountable couplers. One further, and not the least significant, advantage of this type of connector is that it is in some regard tolerant of dirt in that a small piece of grit will only obscure part of the beam and not all of it.
4. *Ferrule type.* Ferrule connectors are quite simple, in that the two fibres to be joined are cemented in place inside small-bore ferrules and the ends polished. The two ends are then inserted inside a third ferrule until the fibres essentially abut, with some mechanism (pads or springs) maintaining a small separation. The whole assembly is then locked or cemented in place and some environmental protection by way of sleeving applied.
This technique is quite good for multimode fibre and can be demonstrated to work for monomode fibres. Depending on the precise design of the connector it can be demountable or permanent.
5. *Kinematic mount type.* This type of connector relies on the geometry of the connecting process to reliably and accurately position the fibre ends. Two geometries are quite common.
The first geometry is where the fibre, or fibres in ferrules, are positioned between

three coaxially aligned bars or rollers to achieve a high degree of precision (depending on how well the rollers are manufactured).

The second geometry is that of interlocking balls or spheres. If you pick up three tennis balls in one hand and a further three in the other then you will be able to rotate these 'triangles' until they fit neatly into each other. So six precision spheres fitted three each to the ends of two fibres for connection will reliably and accurately mate together. Again, with the correct design and manufacture a reliable and repeatable demountable connector can be made (see Figure 7.10).

The above connectors all really refer to the higher-performance mono- and multimode fibres. Fibre systems such as PCS and the larger-core step index fibres do not need the same level of sophistication; this as a consequence of their poor attenuation characteristics. It simply does not matter if the connector loses 3dB if the fibre is losing 150 dB per km. As stated elsewhere in this section, there is a very wide range of simple snap-together connectors for such PCS and plastic–plastic systems.

At the other end of the scale are the military and high-technology cable connectors that interface both electrical and optical circuits and can actually withstand being run over by a tank.

A final form of connector to be mentioned is the repeater station used for very long-haul communications, for example under the sea. Such a 'connector' will receive the weak optical signal, convert it back to an electrical pulse, clean and reshape it, then

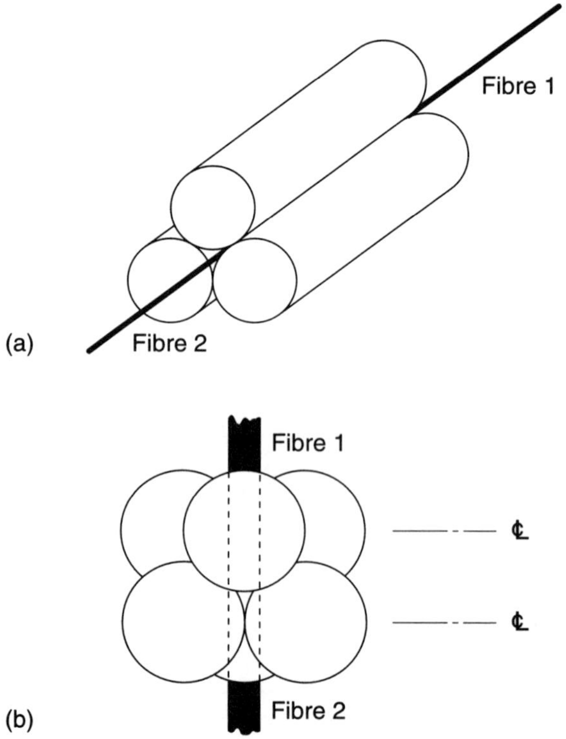

Figure 7.10 Two examples of kinematic connectors: (a) axially aligned bars, (b) interlocking balls

transmit it again as an optical signal to undergo the same process some tens of kilometres later. However, there is intense work under way at present aimed at doing away with the optical-to-electrical and back-to-optical conversions. The reliable in-line and fibre-based communications laser is very nearly with us!

Gradient index (GRIN) lenses

A GRIN lens is one in which the refractive index decreases across its radius or along its axis. Consequently, it has the properties of a normal spherical lens, except that the output faces are plane and parallel to each other. The effect of this refractive index profile on a single light ray can be seen in Figure 7.11 for the four most common types of GRIN lens. In each case, the lens gradually bends the light until it exits the other face in a manner determined by the length or pitch of the lens.

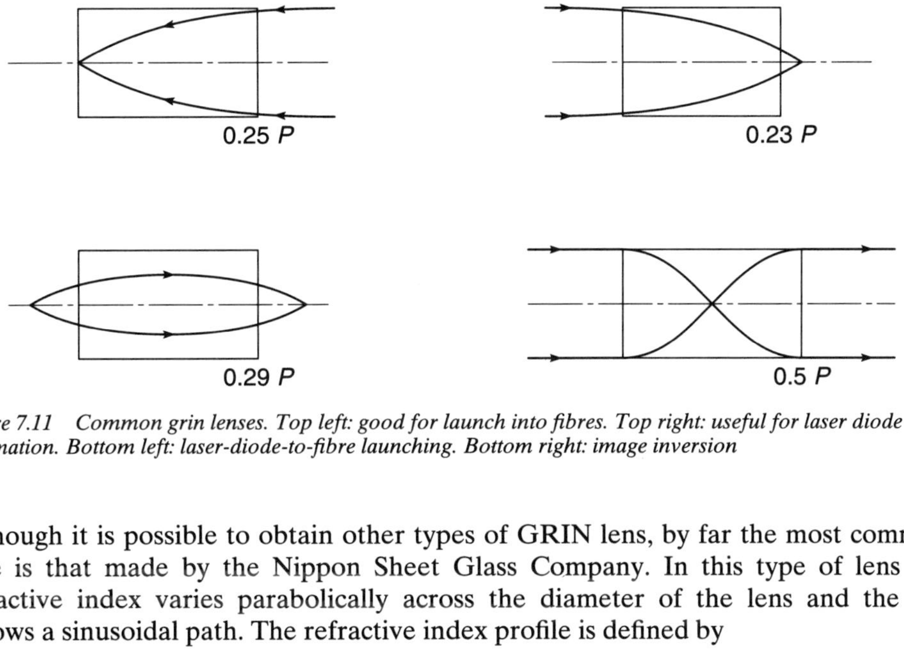

Figure 7.11 Common grin lenses. Top left: good for launch into fibres. Top right: useful for laser diode collimation. Bottom left: laser-diode-to-fibre launching. Bottom right: image inversion

Although it is possible to obtain other types of GRIN lens, by far the most common type is that made by the Nippon Sheet Glass Company. In this type of lens the refractive index varies parabolically across the diameter of the lens and the ray follows a sinusoidal path. The refractive index profile is defined by

$$n(r) = n_{(o)}(1 - 0.5Ar^2) \tag{7.6}$$

where r = radial distance from optical axis
$n(r)$ = refractive index at distance r
$n_{(o)}$ = refractive index at optical axis
A = refractive index gradient constant dependant on wavelength

The pitch referred to above is simply the number of sinusoidal periods completed inside the lens. This can be directly calculated for any GRIN lens by using the formula:

$$P = 2\pi / A^{1/2} \tag{7.7}$$

Typically, the refractive index of the GRIN lens varies from 1.55 to 1.65, and the gradient constant can be anything from 0.3 to 0.6. GRIN lenses can be designed for any pitch or wavelength from the visible to 1550 nm and transmit light as low as 320 nm. But be warned: a GRIN lens designed for one wavelength will not have the same pitch at another wavelength. For this reason, pitches of 0.23 and 0.29 may be better suited to more general applications, although they are most often used in coupling light from laser diodes into fibres.

Obviously, a 0.25 pitch GRIN lens is often used in expanded-beam fibre couplers, especially when it is necessary to have access to the expanded beam – necessary for example in the design of intensity-based fibre sensors (see Fibre sensors). Expanded couplers of this type can still provide insertion losses of 3 dB with a separation of up to 40 to 50 mm.

A 0.5-pitch GRIN lens can also be used as a coupler – often directly between fibres, at a termination or through a bulkhead.

In all cases it is important to match the NA of the fibre with that of the GRIN lens. The NA of GRIN lens is given by

$$\mathrm{NA} = n_{(0)} r A^{1/2} \tag{7.8}$$

and can typically vary from 0.36 to 0.588.

Most adequate optical design software can handle GRIN lenses, provided that the essential lens parameters can be obtained from a supplier or manufacturer.

More detailed treatments for ray tracing with GRIN lenses can be obtained from the first three books listed in Further reading.

Fibre preparation

Glass fibres

Unless preterminated, usually in a ferrule, nearly all glass fibres need some degree of preparation before use. Usually, the plastic outer jacket must be removed, along with any plastic cladding. Then the fibre can be cleaved to give an end that is free from jagged edges and perpendicular to the optical axis.

Several companies market a fibre-stripping tool that pulls the plastic from the fibre, provided it has a plastic cladding. Once the fibre is stripped, commercial cleaving tools can be used to cut it. This is usually done automatically by clamping the fibre, scoring it with a diamond cutting tip and then stretching it over a gentle radius. Although these tools are very reliable, it is often worth checking the quality of the fibre end under a microscope if the application is critical.

Using standard tools for fibre preparation is simple, provided that the fibre diameter matches the range covered by the cleaving tool. If this is not the case, or you don't have a fibre cleaver at all, do not despair; there are cheaper alternatives that can be used.

Stripping the plastic outer jacket or inner cladding can sometimes be done by using a commercial paint-stripping gel.

Some special fibres have a very thin acrylic outer coating that cannot be stripped by any conventional method. In fact, however, the coating can be burned with a match. The burnt residue is then pulled off and the end cleaned with industrial alcohol. This method offers an unusual benefit in that the stripped end is prone to self-cleaving if the fibre is pulled during removal of the residue. The resulting fibre end is often good enough to use without further preparation!

Finally, if a standard cleaving tool cannot be used then the fibre end can be prepared by exposing the bare fibre and applying a single scratch to the outside of the glass with a sharp scalpel blade. The fibre is then placed over a radius (optical mounting posts are useful) and gently bent until the scratch begins to propagate across the fibre and it cleaves. This method requires some degree of practice; the fibre end produced is rarely of the highest quality, but it is often good enough for laboratory or development work.

A quick method of assessing end quality without using a microscope is to launch light into the fibre and view the emitted light through the newly prepared end. It soon becomes obvious if the fibre end is not good enough: the light will be spread out in all manner of directions instead of a well-defined circular beam.

Plastic fibres

Generally, plastic fibres are easier to prepare than glass fibres because the material is softer. Fibre end preparation is equally critical, but is usually done either with a hot knife or by polishing.

Hot-knife cutting uses a standard soldering iron, with the tip replaced by an adaptor holding a scalpel blade. The soldering iron heats up the scalpel and the hot blade can slice through the fibre with very little effort. This method is fast and simple, but polishing may still be required to get a high-quality end.

Polishing gives the best end quality. The fibre can be roughly cut and then put into a metal ferrule or a fibre holder. Successive grades of polishing papers (aluminium oxide papers), discs, or even jeweller's rouge, are used, starting with the coarsest and ending up with the finest finish.

Launching light into fibres

Once a good fibre end has been prepared it remains only to get enough light into the fibre or fibre bundle to suit the application. In many telecommunications applications the light source is a laser diode that has already been coupled into an optical fibre and requires no more effort than making up a standard connector. The real challenge occurs in illumination or sensor design, where a relatively large external light source or possibly a gas laser must be coupled into the optical fibre. In this case, the key is to match the numerical aperture and magnification of the launch optic to that of the fibre and its core size. Although not designed for the purpose, microscope objectives are often used to good effect.

Alternatively, GRIN lenses that are commonly used in launching light into fibres could be employed. The same principle applies: the numerical aperture of the GRIN lens must be matched to that of the fibre.

Applications for fibres

Fibre optics are very common as a means of communication, and are applied in many different circumstances; telephone and video transmission, data transmission, both audio and computer, are just a few. Optical fibres are also used extensively in sensors for detecting pressure, temperature, chemical monitoring, electric current, magnetic fields and rotation. At the other extreme of the applications spectrum are decorative fibre optic lights, fibre bundles for illumination and coherent bundles for imaging.

Incoherent bundles

If a collection of fibres are bound together in a random fashion the bundle is termed incoherent. This means that an image at one end of the bundle is completely jumbled up by its passage through the bundle.

Usually incoherent bundles are used simply as a conduit to move light to remote or inaccessible areas. The combination of a flexible cable and a relatively controlled light output are often attractive design features. In addition, the random nature of the light output can be used profitably to 'scramble' an image. This is particularly useful if the light source has an intensity structure that could be imaged by another part of the optical system. For example, the filament of an incandescent lamp might be visible in a high-f-number projector because of its large depth of field. To overcome this, the incoherent bundle can be used as a relay element to create an even illumination at the projection gate, so that filament structure can't be seen in the final image. Although this can work well, you are warned that incoherent bundles are not always completely random.

A bundle can contain many hundreds of fibres, depending on its diameter and the size of the fibre, so it is quite difficult to produce a genuinely random weave from individual fibres. Often the technique used is to take a smaller number of fibres and route them to a specific location at the other end of the bundle, sometimes incorporating a twist or rotation. When this has been repeated for all of the fibres, the ends are enclosed in an epoxy or adhesive. End faces are then cut and polished. The result can be partially coherent, with regions of the bundle end showing some relationship with each other.

The other big problem with all fibre bundles is their efficiency. Launching light into a single fibre depends on matching the imaging optics to the fibre size and NA. Obviously, a bundle contains many fibres, each only able to collect light over a limited angle. Also, there is a limit to the density with which fibres can be packed together, so not all of the light falling on the end of the bundle can be used. Some manufactures claim packing densities of 85 per cent, but 65 per cent seems a safer factor to apply. Worse still, fibres are often broken or damaged during the assembly process. The net result is a relatively poor efficiency, and fibre bundles of all types should only be used when the amount of received light is not crucial. As a rule you should not expect more than 30 per cent to 50 per cent of the light to be transmitted by the bundle.

A final drawback, common to all bundles, is their sensitivity to infrared. The adhesive binding the fibres together often melts at a much lower temperature (300 to 350 °C) than the fibres can withstand. This produces the disconcerting effect of blackening the end of the bundle as the adhesive first melts and then flows over the fibre end. High-temperature adhesives can be specified, but it is better practice to remove the heat from the input beam before it reaches the bundle ends.

Incoherent fibres are most often used in lighting applications, especially where it is advantageous to keep the source remote from the object to be illuminated or where access is limited. A fibre bundle is often used to provide 'cold' light sources. This simply means that the infrared has been removed from the light before being launched into the bundles.

Coherent bundles

If the fibres of an optical bundle are carefully arranged so that the image at one end of the bundle is faithfully reproduced at the other end then it is said to be coherent. Not surprisingly, this type of bundle is expensive to make and should only be specified for imaging applications.

The most common applications for coherent bundles are in endoscopes, or in any other medical instrument used to obtain an image from within the human body with the minimum size of incision.

Communications

Optical fibres were originally developed for telecommunications, and this market still consumes the vast majority of all fibre produced. Fibres have such a large potential bandwidth that they can carry many signals down the same fibre at one time, and so obviously offer significant cost savings to a systems designer. Local area networks (LANs) and cable TV are among the uses now found for optical fibres under the general heading of communications.

Perhaps not surprisingly, the truth about this area of application is that there really is no need to know very much about optics to design a communications fibre link. Light sources and detectors come already fibre-coupled with preterminated ends, while signal modulation and processing can be integrated into these components and standards such as FDDI fixed ready for the system's operation. Choice of fibre depends on light source – 1300 or 1550 nm – and the design problem is reduced to a systems level specification. The real optics in 'optical fibres' is in sensors.

Fibre sensors

Like a communication link, all fibre sensors comprise three elements: a light source, the fibre in some configuration and a detector. Thereafter the similarity ends, because the fibre can induce changes in the optical signal it is carrying and therefore behave like a transducer.

Optical fibre sensors are electrically passive, so normally they are unaffected by external electrical or magnetic interference, an exception is where the fibre is coated to deliberately sensitize the fibre to one of these effects. Their passivity is also an advantage when they have to be used in explosive environments where the risk of ignition from a stray spark could have disastrous consequences.

Fibre sensors use the optical fibre in one of two ways:

1. In *intrinsic* sensors, changes in the properties of the optical fibre are the means for detection; this type includes acoustic hydrophones, thermometers and electromagnetic sensors.
2. Alternatively, the sensor element can be remote to the properties of the fibre, in which case it is considered an *extrinsic* sensor, for example a chemical sensor.

The mechanisms for sensing change in fibre optic sensors can mainly be grouped into one of the following categories.

Intensity sensors

The simplest means of detecting a change in a fibre sensor is to monitor the intensity of the light propagating through it, either in transmission or in reflection.

There are many ways of affecting the amplitude of the light in a fibre. Some sensors utilize losses from 'leaky' modes to change the intensity, while others such as that in Figure 7.12 use microbending structures that act on the fibre to couple light out of the core.

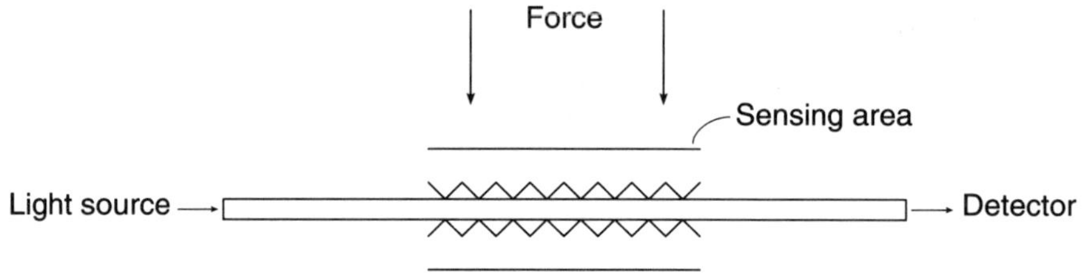

Figure 7.12 Intensity modulation by microbending

Chemical sensors often employ a reflective element at the bottom of the sensor. Light passes twice through the extrinsic sensing element, doubling the sensitivity of the sensor. The components of an expanded-beam fibre coupler can also be used to change the intensity of the light passing from one fibre to another.

Intensity-based sensors are usually used for detecting strain, displacement and sometimes temperature. Their relative simplicity is also their biggest failing, because they are unable to differentiate between intensity changes from the light source or another part of the system and those changes arising from the sensing area itself.

Wavelength sensors

One means of overcoming the problem of intensity changes in the system, rather than those arising at the sensing area, is to use a dual-wavelength approach. In this type of sensor two stabilized light sources of different wavelengths are launched into the fibre. One of the wavelengths is chosen to be unaffected by the sensing element, while the other is intensity-modulated as usual. The result is that a reference channel is now provided, and only differential losses, arising from the sensing area, are detected. An example of this type of sensor uses the Fabry–Pérot cavity to selectively reflect multiples of a discrete wavelength. If the cavity is linked to an external stress then the peak wavelength of the returned light can be monitored without concern for system losses.

Polarization sensors

In this type of sensor an external force introduces a rotation in the plane of polarization of light in the fibre. Usually, low-birefringence fibre is used because it preserves the polarization state of the input light. If the fibre is wound round a current conductor then the polarization state will be altered by the Faraday effect arising from the parallel magnetic field. This type of sensor has been used for current and voltage sensors on high-power transmission lines.

Phase sensors

In Chapter 8 we deal with interferometers and coherence in more detail; it suffices to say at this point that interferometers detect the relative phase between two beams of light from the same coherent source. If one arm of an interferometer is made from optical fibre then the change in phase of the light inside it can be detected. Now the phase of light in a fibre is dependent on the optical path length, which is the product of the length of the fibre, L, and its refractive index, n. If either or both change then the phase of the light in the sensing arm will alter and be detected. Provided that the stress induced in the fibre is not isotropic, both effects work in the same way and enhance the sensitivity of the transducer.

Further improvements in sensitivity can be gained by coating the fibre, using more of it, or winding it onto a former which is sensitive to the force being detected. Plastic coatings, several times the diameter of the fibre, can increase the sensitivity of the bare fibre by ten to forty times. Sensors containing longer lengths of fibre will obviously be more sensitive. So specialist thin fibres can be used, allowing greater length within the same space envelope. Aluminium coatings and segmented formers are used in current sensors.

A typical design for a fibre optic acoustic hydrophone is shown in Figure 7.13. Light from a stabilized laser is divided into two paths; one is launched into the fibre hydrophone and the other forms the reference arm of the interferometer. Often the reference arm is frequency-shifted by modulating the light with a Bragg cell. This makes the sensor a heterodyne system and provides some degree of immunity from low-frequency changes and thermal drift. The acoustically modulated signal from the hydrophone is recombined with the frequency-shifted light from the reference, producing an interference pattern on the photodiode. The output of the detector can be displayed on suitable spectrum analyser or FM-demodulated to yield the original acoustic modulation. The arrangement depicted is a Mach–Zehnder-type interferometer and is common in many sensor designs.

In some sensors, optical fibre is used in the reference path, too. In this case the fibre is wound round a piezo transducer and a control loop built with the detector. Low-

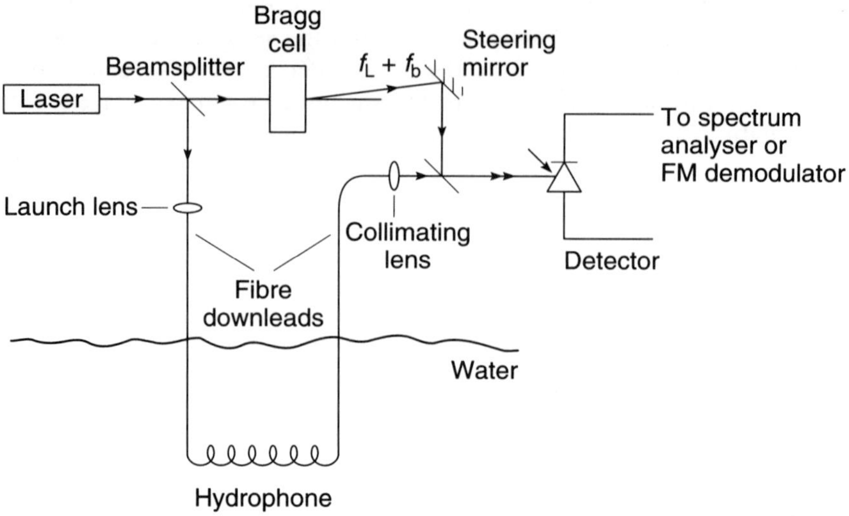

Figure 7.13 Fibre optic hydrophone

frequency changes can be nulled by modulating the length of the reference fibre. Such a system is called homodyne or d.c.

For some time, fibre optic hydrophones have been capable of performing at the minimum detectable pressure in deep sea (62 dB relative to 1 μPa at 500 Hz). Most fibre optic acoustic hydrophones use monomode fibre, but some were built with multimode fibre. The large-diameter fibre made it easy to get efficient light coupling into the core, but the modal pattern produced speckle noise and in the worst case the signal faded out completely. Not surprisingly, multimode fibres have been abandoned for phase detection.

The other area in which phase-dependent fibre optic sensors have been very successful are as gyroscopes. A simplified version of a gyroscope, called a Sagnac interferometer, is shown in Figure 7.14. Light from the laser is divided and launched simultaneously into both ends of a monomode fibre wound round a coil. Some of the light is moving clockwise and some anticlockwise. Rotation of the coil, e.g. clockwise,

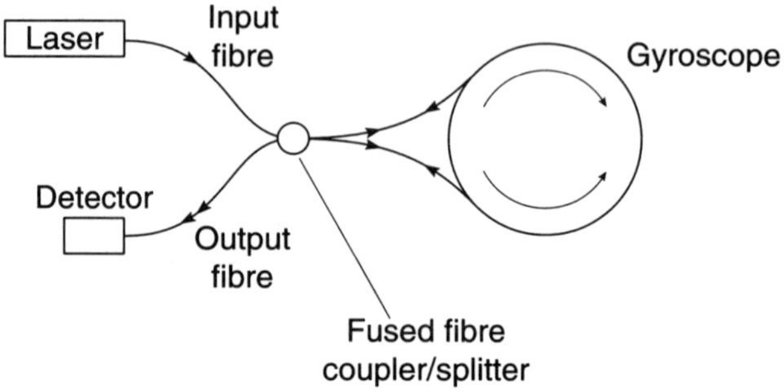

Figure 7.14 Fibre optic gyroscope

shortens the transit time of light in the clockwise path and lengthens it for the anticlockwise path. The two beams are now out of phase when they are recombined and the change can be detected. Although fibre optic gyroscopes are very sensitive, they can be prone to noise effects which affect the relative phase. This problem can be reduced by using polarization maintaining fibre.

This type of sensor is now widely used in the aerospace and even automotive industries.

Mode stripping

Although care is taken to design the optical system so that the majority of the light is launched into the fibre core, some inevitably ends up in the outer casing of the fibre. If this happens in a phase detection system it can lead to noise because the light is not modulated in the same way. To overcome this problem a portion (about 50 mm) of the casing, downstream from the launch point, is stripped away to expose the bare fibre. This region is then surrounded in liquid paraffin to couple light out of the casing but keep it in the core.

CHAPTER 8

Light sources

Introduction

Lasers are now commonplace as industrial tools, as scientific instruments and for providing impressive light shows. In this section the principles of laser operation are introduced together with a discussion of different types and their application.

This is a much simplified discussion of laser physics and laser attributes. Laser physics is a big field, and you can explore the subject in great detail within the literature commercially available. To keep this discussion to the practical point we have avoided much detail; little mention is made of modes, for example. However, the bare bones are here.

The discussion will move on to light emitting diodes (LEDs) once again to cover principles and application. Included here also is a discussion of the semiconductor laser.

Finally the fluorescent lamp and the compact fluorescent will be discussed, with an indication of the different types that are available and the useful spectral range that these can cover.

Lasers

Safety

Let's put laser safety where it belongs – at the beginning. No matter what the power, what the wavelength, what the repetition rate, laser safety is very important. You only have one pair of eyes, so look after them!

When working with lasers you have an obligation to yourself and others to be aware of the national laser safety standards. In the United Kingdom this will be governed by the British Standards Institution (BSI) and in the USA by the American National Standards Institute (ANSI). Many standards are based on the International Electrotechnical Commission (IEC) recommended standard.

Laser hazards come in several guises. These include electrical and optical. Infrared lasers can cause skin damage (burns) and damage to the corneal layers of the eye. Visible lasers can cause retinal damage and ultraviolet lasers can cause eye damage together with skin carcinoma.

Treated with respect, and above all understanding, lasers are not inherently dangerous. We offer here a few simple golden rules in no particular order, as follows:

1. Never, never, never look directly down any laser beam. If you once develop a taste for looking at 'eye-safe' lasers there will come a time when, by habit, you look down a more powerful laser. Don't do it!

2. If you want to see if a visible laser is on or if it works then point it at some diffusely reflecting material and turn it on. Look for the projected spot, not the beam.
3. If you need to observe, say, a small visible laser spot with a microscope or a collimated beam with a telescope then you must exercise a great deal of caution. Use a power meter at the exit pupil of the instrument and use the laser safety standard to determine whether it is safe to look or not. If you are in any doubt whatsoever then do not look down the beam. Neutral-density filters can be used to reduce beam intensity. The best solution is to use a CCD camera, not your eye.
4. When working with an invisible laser beam, use beam probes and power meters to determine exactly where the beam is. Parts of the beam that are unwanted or coming from a specular surface can be physically blocked. With higher-power infrared lasers, firebricks are ideal for this purpose.

Associated with all lasers, except only the low-power semiconductor types, is a high voltage and current power supply. This is in most circumstances more dangerous than the laser beam itself.

Many industrial and higher-power scientific lasers will require the operation of safety interlocks and/or shutters before the laser will operate. In many countries, the USA for example, such interlocks are mandatory for some configurations.

When laser products are designed and manufactured for domestic or public use, safety considerations dictate that every system must be absolutely safe. This includes, of course, a 'fail-safe' condition. If the 'public' system requires a powerful laser then this can be locked and shielded away within the mechanics of the system, so that public access to the beam is absolutely impossible and the total system is safe.

Many laser safety regimes categorize lasers into several different classes. These range from class 1 (eye-safe) through to plain dangerous at class 4. The classification process takes account of wavelength, pulse or continuous wave (CW) and power.

If you are going to work with lasers you are recommended to familiarize yourself with the laser safety regulations and to obtain one of the many recognized texts with respect to laser operation. If you are in any doubt at all then you should contact a laser safety specialist!

Description

The remarkable properties of the laser demand some discussion of how they work. This brief section gives a general idea of how the laser operates and just what makes it so special. As with previous chapters, the 'stimulated' reader is directed to the Further reading list.

Principles

The acronym LASER stands for light amplification by the stimulated emission of radiation. The laser exhibits various properties that are useful to the optical worker.

These are:

- Intensity
- Directionality
- Monochromaticity
- Coherence.

In the following discussions it should become clear how each of these properties contributes uniquely to laser applications.

One concept that we have to understand, or at least accept, before discussing laser action is that each photon has energy proportional to its frequency.

Planck's relationship states that

$$E = h\nu$$

where E = photon energy in joules
h = the Planck constant, equal to 6.626196×10^{-34} J s
ν = oscillation frequency.

Two photons emitted with identical energy have identical frequencies. Two photons emitted with identical frequencies have identical energy.

Using the relationship $c = \nu\lambda$, Planck's relationship can be rewritten as

$$E = hc/\lambda \tag{8.1}$$

where: E = photon energy
h = the Planck constant
c = the velocity of light in a vacuum
λ = wavelength.

So, how does a laser work? Molecules in a gas can both accept and return energy. The energy may be donated to the molecule by heating, collision, absorption of a photon or some other transfer mechanism. The energy can be returned through similar mechanisms.

The amount of energy absorbed or emitted is related to the different energy states or levels of the molecule. The energy difference between two states is fixed and photons released due to downward transitions between these two states will always be of the same energy and hence wavelength ($E = h\nu$).

Under normal conditions the population of states is such that most molecules are of low energy and relatively few are in higher, more energetic states. For laser operation it is necessary to force a population inversion, so that large numbers of electrons exist in a high energy state. Given the correct combination of gases or matter, it is possible to maintain this population inversion by providing energy to the system through some external means.

Stimulated emission is achieved when a photon passing near an excited atom causes a second photon to be emitted. This emission is remarkable, because the additional photon is emitted with identical energy, identical phase and identical direction. Thus coherence and monochromaticity are achieved. The two photons stimulate four photons, the four to eight, the eight to sixteen, and so on. The more material or matter there is to support the laser action, the more intense is the laser source.

The initial, relatively few, photons required to start the stimulated emission process come from spontaneous emission. This natural decay process is responsible for the emission of a few photons almost immediately after the population inversion is established.

Directionality of the photons is organized through the application of two mirrors that form the laser cavity. One mirror is perfectly reflecting and the other a little less so, thus laser light 'leaks' from one end.

In these ways the four laser attributes are achieved: intensity through population inversion and the amount of matter present, directionality, monochromaticity and coherence through stimulated emission.

The above is of course a very simplified discussion. Lasers are available in many different forms, with many singular wavelengths and output powers achieved. However, the principle of stimulated emission and population inversion remains with them all.

Coherence and interference

Lasers are sources of coherent light. This is simply stating that the light from a laser has a fixed phase relationship (this is a consequence of stimulated emission). What the term coherence means in practice can be obtained from consideration of Young's two-slit experiment (see Figures 8.1(a) and 8.1(b)). A narrow slit illuminated by a monochromatic source acts as source to two narrow slits. Each of the two narrow slits serves as a source to illuminate a screen. Consider the point P on the screen which is illuminated by light from both slits A and B. If the light from each of the slits arrives in phase, that is peak to peak and trough to trough, then

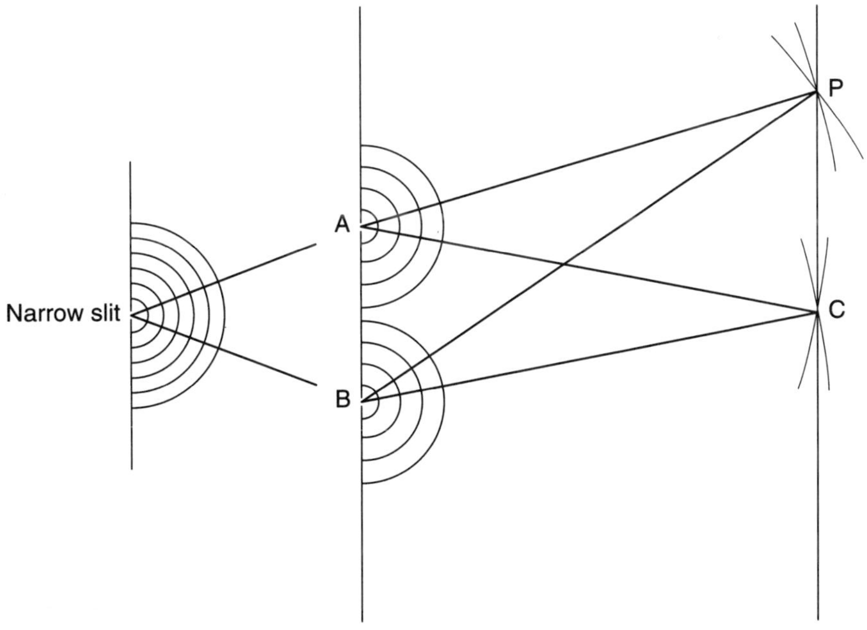

Figure 8.1 (a) Young's two-slit experiment.

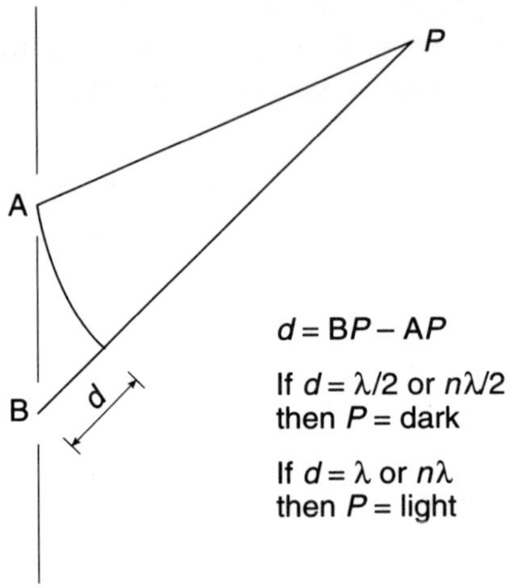

Figure 8.1 (b) Young's two-slit experiment

constructive interference will produce a light stripe. If the two waves arrive exactly out of phase, that is peak to trough, then destructive interference will occur and a dark stripe will be observed.

Let us explain what we mean by interference. Look at Figure 8.2(a) and (b), which

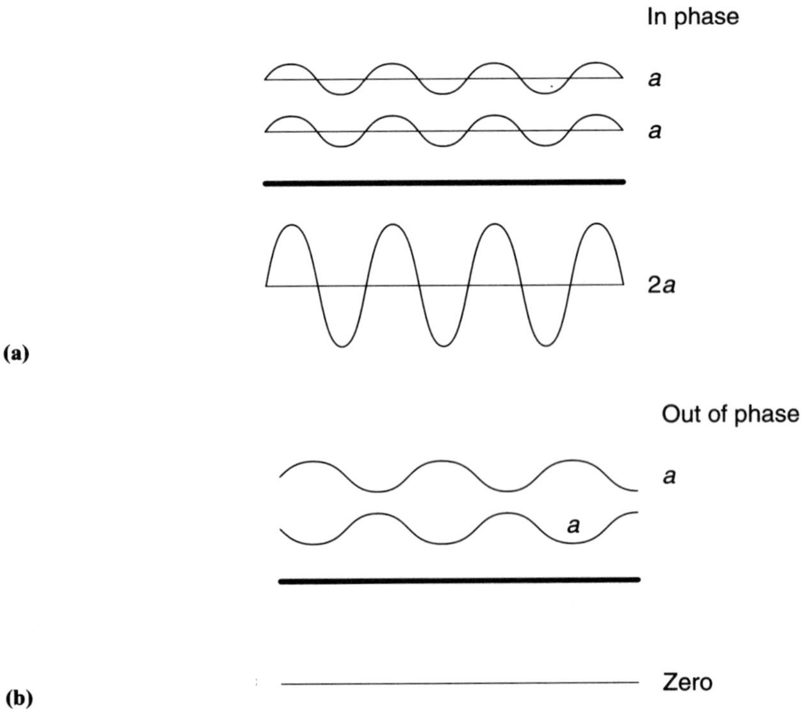

Figure 8.2 (a) Interference and wave addition. (b) Interference and wave subtraction

illustrate the principle of wave addition. When waves travel in phase, peak to peak, then the wave amplitude is additive, e.g. $a + a = 2a$. When the waves travel exactly out of phase then the wave amplitudes cancel $a + (-a) = 0$. This is called respectively constructive and destructive interference. Implied in Figure 8.2(a) and (b) is that the waves are of identical wavelength, that is, the source is monochromatic.

When a point (or narrow slit) source radiates light, the resulting wavefront is of equal phase. This is after all a definition of the wavefront. So, with Young's experiment described above, the small size of the primary source dictates that the light is emitted essentially from the same point; thus the light reaching the two slits is of common phase. The subsequent slits act as spatially separated coherent sources emitting light with the same phase. If at some point P on the screen the distance d is an equal number of wavelengths then constructive interference occurs. If, however, d is an odd multiple of half-wavelengths then destructive interference will occur. In this way a pattern of light and dark stripes is generated.

If the light from the two slits was incoherent, perhaps with a different light source for each, with no phase relationship, then no interference effects would be observed. Thus, independent sources are incoherent and have no fixed phase relationship. This is why two independent lasers of the same type cannot be made to interfere: they have a different phase relationship.

Coming back to the original point; stimulated emission yields light with a fixed phase relationship; thus laser emissions are coherent.

Two more terms you will have to become familiar with as an optical worker with an interest in lasers are coherence length and bandwidth. They have a simple relationship with each other: the longer the coherence length, the narrower the bandwidth. What do we mean by these terms? Let us take coherence length first.

The coherence length of a laser is really the physical distance over which the pencil of rays emitted from the laser travel with the same fixed phase relationship.

A low-pressure mercury lamp may have a coherence length of a few millimetres or so, whereas a simple gas laser may have a coherence length of 200 mm to 300 mm and a well-constructed and stabilized laser one of more than 100 m. The fact that coherence lengths are not infinite indicates that lasers are not truly monochromatic. Each laser line has a bandwidth, or line width, which is tightly related to coherence length and time.

As a general rule of thumb the coherence length L is related to the bandwidth $\delta\nu$ by

$$L = c/\delta\nu \tag{8.2}$$

where c is the velocity of light in a vacuum.

By way of example consider two different lasers, a helium–neon gas laser and a representative semiconductor laser. The He–Ne laser may have a bandwidth, full width at half maximum (FWHM), of about 1500 MHz, which gives a coherence length of the order

$$L = 3 \times 10^8/1500 \times 10^6 = 0.2\,\text{m} \quad \text{or} \quad 200\,\text{mm}$$

A short-wavelength semiconductor laser will have an instantaneous linewidth that varies with how the device is operated; just above threshold the linewidth could be of the order of 4 nm, while well above threshold it might be around 0.1 nm.

Assuming an average linewidth of 2 nm, this is converted to bandwidth by the following rote:

The linewidth equals ±1 nm; if the centre is, say, at 780 nm then we are interested in the interval between 779 and 781 nm. The frequency of each, from $c=\nu\lambda$, is given by

$$\nu_{779} = 3\times 10^{8}/779\times 10^{-9} = 3.8511\times 10^{14}$$

and

$$\nu_{781} = 3\times 10^{8}/781\times 10^{-9} = 3.8412\times 10^{14}$$

$$\delta\nu = 9.8619\times 10^{11} \quad \text{(i.e. 986.19 GHz)}$$

and

$$L = c/\delta\nu = 0.3\,\text{mm}$$

This very simple analysis illustrates the great difference between coherence lengths for standard gas and semiconductor lasers. However, some specialized semiconductor lasers are capable of much better coherence lengths of the order of a metre or so.

One test and measurement application where coherence length is important is that of interferometry. If the arms of, say, a Twyman–Green interferometer are not well matched (so that there are not similar optical path lengths through each of the arms) then it will be extremely difficult to obtain fringes of good contrast. However, a laser with long coherence length can tolerate a difference in path length of tens of metres; hence the emergence of the laser unequal path interferometer (LUPI).

The relatively short coherence length of some semiconductor lasers can be of advantage in certain applications. For example, a coherent reflection from the outer surface of a compact disc could interfere with the reflected signal and produce a null.

There is one further phenomenon to discuss with respect to coherence, namely that of speckle. A laser beam when viewed in reflection from a slightly rough surface will appear to take on a grainy appearance or to shimmer. This effect, which will vary with the slightest movement of the observer, is called laser speckle.

Speckle manifests itself through interference due to the slightly different path, and hence phase, differences that arise from reflection from the peaks and troughs of the surface.

Some relevant laser properties

Lasers and laser beams are characterized by a number of properties that vary in importance according to the application. Some of the most common parameters are discussed below. It is important to understand these parameters and their relation with each other. Fundamentally, there are some things you can do with a laser and some you can't. The fundamental issues that control the application of a laser are described below.

Modes

In this context, mode really means intensity distribution across the laser beam. The majority of commercially available lasers operate in what is referred to as the TEM_{00} mode. The 'TEM' part of the specification refers to the propagation direction of the beam which is orthogonal to the transverse electric and magnetic fields; The '00' part designates the simplest transverse mode of the laser, which is propagation with a centralized Gaussian intensity distribution. There are other, more complex modes of operation, but unless you are about to get involved with high-power lasers, or lasers with interesting cavity designs, you will not encounter them.

A higher mode is usually manifested as a 'blobby' beam with several intensity maxima. Beams can be near-square or rectangular, so, for practical applications that involve the use of lasers with higher-order modes, you may have to consider beam-shaping optics to mould the beam to a form that can be controlled by conventional optics.

Beam waist and divergence

A natural laser beam does not have a sharp edge of high contrast. Peak intensity is usually at the beam centre, with a gradual falling away of intensity toward the edge. For most gas lasers this beam intensity profile is Gaussian. The diameter of a Gaussian beam is usually defined as that diameter where the beam intensity falls to $1/e^2$ of the centre intensity. The beam waist, which is the natural focus of the laser beam and is a function of cavity design, usually occurs at the output mirror or aperture of the laser.

The beam waist is usually extended in space, and can be very small for highly converging beams or large for slowly converging beams. However, the waist marks the transition from a converging beam to a diverging one.

By definition, then, somewhere within the beam waist the wavefront must be plane, not spherical, and it is this point that defines the location of the waist. The diverging wavefront will have a radius of curvature $R(z)$ and a beam radius $r(z)$, after travelling a distance z from this point.

Laser beams cannot be perfectly collimated, and in the limit the divergence of the beam is fixed by diffraction at the exit aperture. Such a laser is said to be diffraction-limited, which is the case for nearly all lasers operating in the TEM_{00} mode. Lasers that operate in other than the TEM_{00} mode may have complex cavity designs that add aberration to the wavefront, which will add to the divergence caused by diffraction.

Most lasers, except semiconductor types, will exhibit a very low beam divergence of less than 10 milliradians (mrad), and usually one that is nearer 1 to 2 mrad. The divergence angle quoted by suppliers normally refers to the full angle of the beam – but check the specification to make sure!

The beam spread due to diffraction can be predicted directly from diffraction theory, and on the whole agrees well with practice. At a great enough distance from the beam waist the divergence is given by

$$\text{divergence} = 1.22\lambda/D \tag{8.3}$$

where D is the waist diameter and λ is the wavelength.

For an He–Ne laser with a wavelength of 632.8 nm and a beam waist (diameter) of 0.8 mm we would achieve a steady-state divergence of

$$(1.22 \times 632.8 \times 10^{-9})/0.8 \times 10^{-3} = 1\,\text{mrad}$$

Beam spread can be decreased by moving to a shorter wavelength or to a larger output aperture. For example, doubling the beam diameter above results in a divergence of 0.5 mrad.

A little more detail will be useful here for anyone involved with the practical application of a laser. The behaviour of a laser beam near focus is not described by the divergence equation quoted above. This is why we added the caveat 'when sufficiently far from the beam waist'. You may be interested in what happens close to the laser beam waist, in which case you should note the following:

$$R(z) = z\,[1 + (\pi r_0^2/\lambda z)^2] \tag{8.4}$$

and

$$r(z) = r_0\,[1 + (\lambda z/\pi r_0^2)^2]^{1/2} \tag{8.5}$$

where $R(z)$ = radius of curvature of the wavefront after travelling a distance z from the waist or point where the wavefront was plane
z = distance travelled from the waist
r_0 = beam waist radius ($1/e^2$ point)
λ = wavelength
$r(z)$ = radius of the laser beam ($1/e^2$ point) after travelling a distance z.

Consider the second of these two equations for the case when z is much larger than $\pi r_0^2/\lambda$. In this case we can write

$$r(z) = \lambda z/\pi r_0 \tag{8.6}$$

The diverging laser beam can be considered as a cone with its apex at the beam waist. The full cone angle will be given by

$$2r(z)/z = 2\lambda/\pi r_0 \tag{8.7}$$

For practical purposes the simplified form of $r(z)$ can be used when $z \geqslant 10\pi r_0^2/\lambda$.

From these equations the beam diameter and the curvature of the wavefront can be identified at any point within your optical system. This enables identification of aperture sizes and component diameters, and generally allows an accurate description of the system.

For a simple visualization of how these parameters fit together see Figure 8.3.

Polarization

Lasers can be linearly polarized or randomly polarized. In the latter case the laser emits a truly random combination of linear polarization states the composition of which can vary with time. Beware: use of a randomly polarized laser within a polarization-sensitive application can give results that vary with time!

In the case of linearly polarized lasers, the mechanisms that yield different polarization states are suppressed, and Brewster windows ensure that only the wanted polarization state is emitted; the linearity is unwavering.

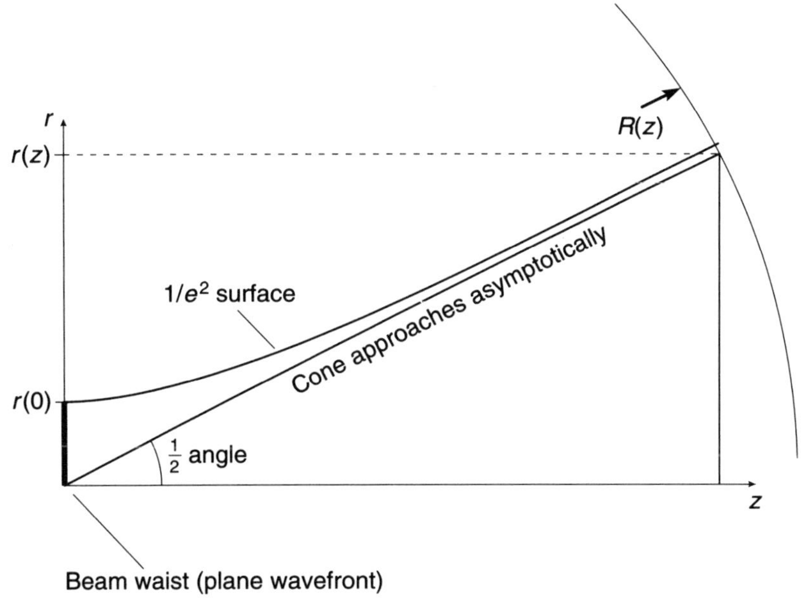

Figure 8.3 *Laser beam waist and divergence*

Focusing a laser beam

One result of the excellent spatial coherence of a laser source is that the ray bundle can be focused to a very small spot, which is useful, for instance, with high-power-density applications.

To understand the difference between focusing a spatially coherent laser beam and focusing a spatially incoherent source such as a white light illuminated diffuser, consider Figure 8.4. Part (a) shows the collimated laser beam with each part of the beam carrying a fixed relationship, with every other part being brought to a very small focus; For a well-designed laser-focusing lens the spot diameter so produced will be of the order of the diffraction limit. Part (b), on the other hand, shows that a spatially incoherent source is not brought to a small focus. Classic ray optics show that the source aperture is simply reimaged.

The well-designed laser-focusing lens will be optimized for:

1. The specific laser wavelength.
2. The laser beam diameter.
3. Minimum aberrations (particularly spherical).

To get the very best from a best-form laser-focusing lens it has to be used correctly. When you consider the production of a very small diameter laser spot, there exists a subtle interplay between the effects of geometry and diffraction, with the result that the lenses are usually specified for a maximum laser beam diameter. As always, the steeper side of the lens should be directed toward the focus.

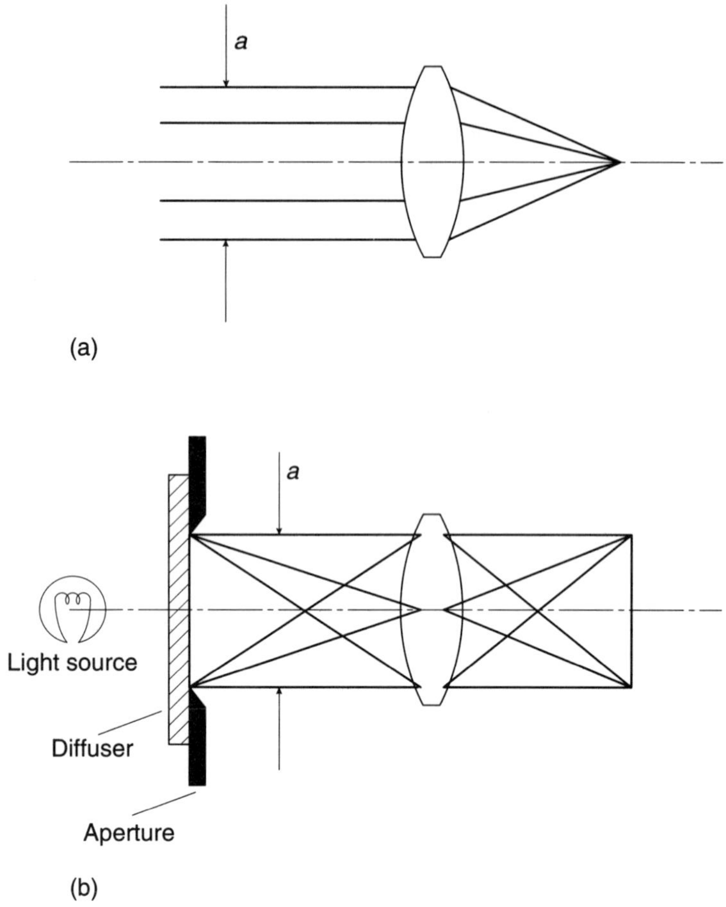

Figure 8.4 Focusing a laser beam. (a) Laser: spatially collimated ray bundle. (b) Diffuse: the incoherent source is reimaged and not brought to a small focus as with the coherent laser source

Expanding a laser beam

We have already discussed the fact that a laser beam has a natural divergence angle. This can be modified through expansion or compression of the beam. If the laser beam is expanded then the divergence will fall, which, for example, can be useful for transmission through safety tubes or for any application where good beam control is required. Conversely, if the beam is compressed then divergence will increase. A general rule that can be applied is

$$\text{beam divergence (mrad)} \times \text{beam diameter (mm)} = \text{constant}$$

The magnitude of the constant varies between laser types, but for a fairly standard He–Ne it is about 0.81.

How to specify a laser

As with virtually any piece of scientific equipment, and as a general rule throughout commerce, you get what you pay for. Highly stabilized and rugged lasers operating at wavelengths found only within a few applications will cost far more than the ubiquitous laboratory He–Ne.

Most importantly, as a laser user you must get completely clear in your own mind exactly what you require from the laser system you wish to purchase. Parameters to be considered include:

1. Pulsed or CW
2. Single-wavelength or multi-spectral
3. Absolute power requirements
4. Output stability
5. Safety
6. Monochromaticity
7. Optical component implications
8. Required power supplies and other services
9. Cost
10. Integral power supply or head-only
11. Cooling method
12. Warm-up time

Commercially available lasers will be specified in several ways, and different manufacturers will put their own emphasis on what is and what is not important. Always bear in mind the use to which the laser will be put. Particular parameters will include those shown in Table 8.1.

Table 8.1 Specification bases of commercial lasers

Parameter	*Unit or class*
Type	—
Power output	mW
Wavelength	nm
Mode	Usually TEM_{00}
Polarization	Random or linear
Beam diameter at $1/e^2$ points	mm
Beam divergence	mrad
Time to 95% full power	min
Power supply requirements	mA
Laser safety class	I, II, III, IV
Space envelope	mm
Weight	g
Linewidth or longitudinal mode spacing	MHz

Laser types

In the following tables (8.2 to 8.4) some common laser types are introduced, together with their salient characteristics. The lists are not exhaustive, but should give an insight to the major types, their differences and application. Again there is a wealth of excellent literature, which gives a far more detailed picture than that presented here.

Semiconductor lasers have been deliberately omitted from this discussion because they have their own section below.

Ultraviolet lasers

Table 8.2 Ultraviolet lasers

Type	*Wavelength*	*Characteristics*
Argon ion	275 to 306 nm	CW devices with output power of up to 1 W
He–Cd	325 nm	CW, tens of milliwatts
Nitrogen	337 nm	Pulsed, under 1 W average

The common types are shown in Table 8.2; besides these are those in the excimer range of lasers. An excimer laser belongs to a family of lasers that are particularly useful in the ultraviolet region of the spectrum. They are characterized by short, energetic pulses.

Visible lasers

Table 8.3 Visible lasers

Type	*Wavelength nm*	*Characteristics*
Argon ion	450 to 520	CW, tens of watts
Copper vapour	510, 578	Pulsed, tens of watts
He–Ne	632.8	CW, 50 mW
He–Cd	442	CW, 50 mW
Ruby	694	Pulsed, 1 or 2 W
Dye	370 to 1000	CW, a few watts

Among these, tunable dye lasers tend to be used for specialist applications where very short pulses or very narrow bandwidths are required. They are inefficient and tend to be pumped either by flashlamp or by external laser.

Infrared lasers

Table 8.4 Infrared lasers

Type	*Wavelength μm*	*Characteristics*
Nd–YAG	1.064	CW or pulsed, kilowatts
CO_2	9 to 11	CW or pulsed to tens of kilowatts

Laser applications

Lasers are found in many applications, and a substantially incomplete list would include the following:

- Telecommunications
- Holography
- Compact disc, video and computer
- Exciting light shows
- Rangefinding
- Materials cutting and welding
- Product or material gauging
- Bar code reading
- Target designation
- Surgery

This list could go on and on. However, it does show that from a hesitant start laser technology is now fully developed and mature. To complete the picture a few of the applications above are discussed briefly below.

With respect to measurement systems, the laser offers the system designer an opportunity to reject ambient light from the measurement cell. A narrow bandpass filter centred on the laser wavelength will serve to reject the vast majority of ambient light, thus easing the detection task.

Helium–cadmium lasers

The helium–cadmium laser provides a useful source of deep blue, near ultra violet light or both simultaneously. Two lines are of most interest, these are 441.6 nm where low hundreds of milliwatts are available and at 325 nm where powers of tens of milliwatts can be achieved. The lasers are supplied as sealed units and the more powerful versions may require a degree of forced air cooling.

Applications are various but are mostly concerned with information technology or inspection. The shorter wavelength of the helium–cadmium means that focused spot sizes are small, useful in high density printing or optical recording, for example. The same small spot size can be used as a probe in measurement or inspection applications.

Some care has to be exercised when selecting appropriate optical systems or lenses. Most optical glasses will retain a usable internal transmittance at 442 nm, however, at 325 nm most glasses will be absorbing to a greater or lesser extent and special UV grade glasses should be sought.

Some characteristics of the laser are worth mentioning. In order to maintain CW operation the tube needs to be heated to retain the cadmium at the required pressure. Temperatures of about 250 °C are required. Beam quality is good and highly polarized outputs are possible.

The He–Cd laser emits at visible wavelengths, and hence all of the usual safety rules apply. The UV emissions from an He–Cd can reach the retina but are mostly absorbed at the cornea, and it has been reported that UVA can contribute to the formation of cataracts.

Helium–neon lasers

These are probably the most common and familiar of all lasers. But, for how long will these particular gas lasers hold pole position? The semiconductor lasers are dropping fast in both wavelength and price, and offer advantages such as smaller size, fast direct modulation, mechanical ruggedness and the absence of any requirement for high drive voltages. However, the helium–neon laser does have a better beam quality and longer coherence length, so it is unlikely to be replaced in the near future.

The many uses of He–Ne lasers include laser demonstrations, laboratory work, measurement, some holography and so on. Bearing in mind our comments above with respect to semiconductor lasers, we highlight below those areas of application where the He–Ne excels and seems unlikely to be replaced very easily.

Also available now is the green He–Ne laser. This operates at a wavelength of 543.5 nm, which falls near the peak response of the eye, so it might find good application as a pointer or indeed in any application where good visibility is required. Orange and yellow have also become available in recent years.

Measurement

For precision displacement systems that count interferometer fringes produced over distances of a metre or so, a stabilized He–Ne is ideal. Low cost, long life (greater than 20 000 hours, which is over 2 years' continuous operation!) and long coherence length are relevant parameters.

Interferometry

The long coherence length of the He–Ne laser makes it a good choice for interferometric applications. Simple interferometers can be used for the optical quality assessment of lenses, optical flats and wedges, and mirrors. Many practical and commercially available, workshop-quality test tools are based on such systems. Advanced interferometry can find application for the He–Ne in such fields as Doppler-shifted interference patterns for the remote measurement of velocity.

Holography

Holography can be regarded as an interferometric application. The recording of patterns from the interference of coherent reference and scattered measurement beams on a high-resolution film is possible only if the coherence length of the source is long. Again, the He–Ne laser is a relatively cheap, reliable and visible source of such radiation.

Nd–YAG lasers

The Nd–YAG (neodymium–yttrium aluminium garnet) is perhaps the most familiar of the neodymium family of lasers. It is a solid-state laser (i.e. constructed with a solid rod, not a semiconductor!), with external mirrors used to create the laser cavity.

Excitation can be achieved through a variety of mechanisms, which include other lasers (semiconductor), flashlamp or continuous arc. Generally the laser is simple and versatile, and produces a good-quality, continuous beam at room temperature.

The Nd–YAG operates at a wavelength of 1.064 μm and can deliver a high power in either pulsed or CW mode.

This laser represents a very real eye hazard. The cornea and lens are effectively transparent to this wavelength, and hence the laser output can be focused to a small spot at the retina. The high power density of the small spot invariably causes damage and, as this type of laser is usually of a high power, permanent loss of sight. The laser light is invisible and hence the observer has no blink aversion reflex. Laser safety goggles are essential!

Owing to the eye lens's capability to transmit light of this wavelength, the Nd–YAG can be used for laser eye surgery (cutting or welding of the retina). Traditional fibre optic cable will transmit well at wavelengths about 1 μm, which results in some medical–fibre-optic applications.

The Nd–YAG laser can be focused to a very small spot size, which enables such industrial applications as resistor trimming and pulsed-mode spot welding and drilling.

Military applications include laser target designation and rangefinding. The high powers available within a transmission window of the atmosphere make the Nd–YAG a natural choice for these applications.

Carbon dioxide lasers

The CO_2 laser is the most common infrared, or thermal, type, operating at wavelengths of between 9 and 11 μm. High-power, continuous-wave outputs are available and make this laser particularly suitable for industrial applications such as welding, cutting and marking.

The CO_2 laser is versatile and a few distinctly different types are available, including the following:

Waveguide

This type is stimulated through intense radio frequencies or d.c. discharge. The laser tube is reduced in size so as to form a waveguide structure. This type of CO_2 laser is compact and has a good beam quality. Output is limited by difficulties in removing enough heat from the process so as to maintain an efficient laser action.

Sealed tube

Here the CO_2 gas mixture is sealed inside the laser tube. Compared with flowing gas types (discussed below), this type is characterized by relatively low powers, again because of heat removal problems. Water cooling is not unusual.

Flowing gas

Such lasers are characterized by high powers. The gas flow provides a degree of cooling, with an advantageous gas condition to maintain efficiency. The highest-output versions have fast gas flow and water cooling. This type is a common industrial tool.

With any CO_2 laser, safety is again an issue, because of the high-power invisible beam. The human eye will not transmit at these wavelengths, but corneal burns are very unpleasant. Besides such eye hazards there exists that of skin burns. Be aware of where the beam goes, and always control potential reflections.

Because of the high thermal power of a CO_2 laser, the optical system is usually constructed from reflective optics only. High reflectivities are essential, together with active cooling to prevent damage. A 0.1 per cent absorption at a reflecting surface of a 25 kW laser beam can result in significant power being absorbed by the optical component. Focusing can be achieved by elliptic or off-axis parabolic components. These are likely to be manufactured on a water-cooled substrate and coated with gold to maintain the highest reflectivity at 10.6 μm.

Applications of CO_2 lasers are numerous and include:

- Interferometric testing of thermal imaging lenses.
- Welding, marking and cutting of even highly reflective materials.
- Heat treatment.
- Surgery.

Frequency doubling

Frequency doubling is an optical process in which the frequency of an optical beam is doubled, and hence the wavelength halved, in a coherent fashion.

This technique can be applied with good effect to produce visible light from near-infrared wavelengths or ultraviolet from visible laser sources. For example, a frequency-doubled Nd–YAG laser will produce radiation of 532 nm from the natural 1.064 μm emission.

The principle of energy conservation implies that as the frequency is doubled, and the wavelength halved, energy should double; this is implied by the formula $E = h\nu$, where E is the energy, h is the Planck constant and ν is the frequency. However, the mechanism is more akin to two weak, long-wavelength photons combining to produce one strong, short-wavelength photon. Conversion efficiency is hence necessarily low. An efficiency of about 20 per cent is usual.

The frequency-doubling effect is achieved through the non-linear interaction of an electromagnetic wave and the material through which it is passing. Many non-linear crystals are available, the most common being KDP, $LiIO_3$ and $LiNbO_3$; which crystal type is selected for the application will really depend on the operating wavelength and the optical power applied.

The frequency doubling is part of a harmonic generation effect in which the second, third, fourth and much higher harmonics can be generated.

By way of application, diode-pumped, solid-state lasers together with frequency-doubling crystals can produce relatively high-power levels of green light with a good beam characteristic. These devices are more expensive than, say, an ion laser, but do offer the advantage of compactness.

Light-emitting diodes

The light-emitting diode (LED) is basically a classic diode as used by electrical engineers throughout the world. However, instead of generating heat at the forward-biased p-n junction the recombining carriers emit light. In short, the LED emits light when biased in the forward direction.

The actual amount and the characteristics of the light thus emitted depend on the types of materials used to manufacture the p-n junction and on the current applied. The material, together with its impurities and doping levels, which control bandgap energy, controls wavelength and the electrical current controls brightness. The light emission is spontaneous, not stimulated as with laser devices.

LEDs have relatively narrow bandwidths, typically of the order of ±25 nm. Many output wavelengths are available, with more towards the red end of the visible spectrum than towards the blue.

Devices may be driven continuously across a fairly broad range of forward currents, or alternatively may be pulsed. Typical response times are of the order of tens of nanoseconds.

Most visible diodes are specified in terms of millicandelas (mcd) at a particular, rated, forward-drive current. Many LEDs are available with lenses ready-moulded on the front. A lens that provides a relatively narrow beam of light will provide a higher output (in millicandelas) than the same LED without a lens. The lens controls mcd output for a fixed lumen value.

Light-emitting diodes are available through the visible and near-infrared spectrums. More accurately, an infrared-emitting diode should be called an IRED, but in most contexts today LED is used for the whole of the spectrum.

Some relevant LED properties

When you come to consider the specification of an LED for your particular application there will be an initial list of attributes to satisfy. Once some potential diodes have been selected, a more detailed specification will be formed.

The first specification will include such parameters as:

1. Colour of output (wavelength)
2. Package
3. Lensed, tinted or diffused
4. Luminous intensity
5. Emission angle
6. Typical forward voltage

Once a device has been selected that meets the initial requirements, determined from the list above, the system designer will move on to determine:

7. Power dissipation
8. Thermal characteristics
9. Peak forward operating current
10. Reverse voltage

11. Resistance to transients
12. Operating and storage temperatures
13. Soldering characteristics
14. Lifetime
15. Continuity of supply
16. Cost!
17. Construction
18. Rise time

These characteristics are discussed below.

Colour of output

Much effort is being expended by the manufacturers to produce the most efficient blue LED. The fledgling devices commercially available represent the lowest wavelengths currently achievable. The rest of the visible spectrum is well catered for with green, yellow, orange and red devices available. Some 'white' LEDs are now becoming available that comprise, essentially, a red, a green and a blue LED in a single package.

Devices that emit in the infrared are available, with particular use in communications where advantage can be taken of fibre optic transmission windows. For example, a GaAlAs LED emitting at a wavelength of 850 nm is well placed to take advantage of plastic-clad silica (PCS) fibres with a transmission window of between 800 and 880 nm.

Package

LEDs can come in a variety of packaging styles. These styles range from the classic transistor can (in 'T' sizes, such as T05 for example) to the flat surface emitter, from a 'blob' on legs to a packaged device ready to receive a fibre optic. There are also various lead configurations.

Typically, a simple LED will be cylindrical with either a lensed or a flat at one end and two legs – anode and cathode connections – at the other. Usually, packages will be labelled T-1, T-1 3/4 or T2, which means 3 mm, 5 mm and 6 mm in diameter respectively. Some high-power LEDs use a 10 mm diameter package.

Lensed, tinted or diffused

For a fixed lumen output a lens formed within the LED package will control the mcd output. Given a fixed lumen output, a high mcd figure means a narrow beam, a low one a wide beam.

The LED package itself can be clear or coloured. A clear package has a slightly higher output owing to the lower transmission loss through the material. However, it may be desirable within some applications to hide the internals of the LED itself, and here a tinted package would be more appropriate.

Some packages offer a fairly well controlled beam, such as the lensed-end types, whereas others offer a very diffuse output with little optical control.

Luminous intensity

Luminous intensity relates to the perceived brightness of the LED source; the unit of measurement is the candela (lumen per steradian). For a given luminous power, measured in lumens, the luminous intensity is controlled by the solid angle into which the LED emits.

Some visible LEDs that emit into relatively small solid angles are rated at high luminous intensities compared with those that emit into large solid angles. Some examples are given in Table 8.5, which shows the decrease in mcd rating as the beam width increases for essentially the same device.

Table 8.5 Visible LEDs

Colour	*Lens*	*Luminous intensity mcd*	*FWHM*	*Forward voltage at 20 mA V*
Red (626 nm)	Not tinted	125	24°	2.2
Red (626 nm)	Not tinted	55	32°	2.2
Red (626 nm)	Not tinted	35	45°	2.2

In the table we have not considered wavelength. The mcd rating is related to an average human eye response and as such will increase near the green (peak human eye response) and fall near the red or blue.

It is a fair assumption that when a LED with high luminous output is selected its emission angle will be small. In common with laser safety requirements, high-power LEDs should be treated with great care to prevent eye damage.

Emission angle

This feature of LEDs is very application-dependent. For launching into a fibre optic a small emission angle is required, whereas for general illumination or perhaps hazard lamp illumination (as on a car dash panel) a wide emission angle is required. It is up to the user to decide which emission angle is suitable for any particular application.

Typical forward voltage

The forward voltage is usually of the order of a few volts, within the approximate range of 1.5 to 5.0 volts.

Power dissipation

The power dissipation or consumption of LEDs is generally very small. Consider a small device with a forward current of 10 mA at a few volts: the power consumption will be of the order of 30 mW. The more powerful devices will of course consume more power!

Thermal characteristics

The maximum permissible forward current that can be applied, before the device breaks down, falls as the ambient temperature increases.

The thermal characteristic is again device-dependent, but to make a gross generalization performance at up to 50 to 60 °C should be acceptable. See Figure 8.5 for the general characteristic.

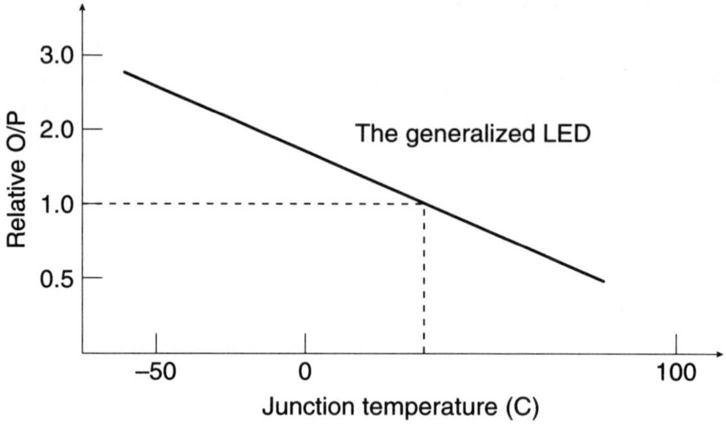

Figure 8.5 Thermal characteristics: LED output with temperature

Peak forward operating current

The efficiency with which an LED emits light both rises and falls with the applied forward operating current. Thus there is a peak forward operating current at which the light emission is at its most efficient. This value will usually be stated by the manufacturer.

The relative efficiency with which light is emitted falls at higher values of forward current because of heat dissipation within the device itself. At and near the optimum the function tends to be relatively flat, so operation near the peak is usually good enough.

Reverse voltage

This occurs when the cathode is positive with respect to the anode, and if the reverse voltage, and current, exceed the specified capability of the diode then it will be destroyed. The maximum specified reverse voltage is in effect the breakdown voltage of the diode.

The reverse voltage and current specifications are really a warning that if the diode is reverse-biased to more than, typically, 5 V at 100 μA then the diode will be damaged. Because it is easy to put an LED into a circuit board the wrong way around, this is quite an important specification.

Most diodes will have the cathode marked in some way. However, a few manufacturers mark the anode, so be sure to check which is marked on your diode.

Resistance to transients

Most semiconductor light sources, especially laser diodes, are very sensitive to transients on the power supply lines. These transients can cause failure of the devices. Therefore care should be taken with the design of power supplies to ensure that transients are minimized.

Operating and storage temperatures

Light-emitting diodes will be specified for operation and/or storage over a particular temperature range. The systems designer must ensure that the selected devices are capable of operation and storage over the range of temperatures required for the system.

Storage temperature is often overlooked, but can be very important. For example, if the equipment being designed has to be air-portable then extremes of low temperature will be experienced in the cargo hold.

Soldering characteristics

High temperatures cause catastrophic failure of LEDs. Most manufacturers will supply a specification of maximum soldering temperature with the maximum time of application of the iron or whatever. A typical specification might be 250 °C for 5 seconds.

Lifetime

LED lifetimes are usually of the order of 10 000 hours for continuous operation. Operation of an LED above its rated values or specification will reduce its life. If the diode is kept within its specified limits for operation then modulation will not significantly affect its lifetime. For low modulation rates lifetime really equates to the time during which the diode actually emits.

Continuity of supply

If your selected LED is very special for some reason, which might include a high output, a non-standard colour, a particular lens, etc., then it will be a good insurance to make sure that a second source of that particular diode is available. Unfortunately it is not uncommon for a product to be redesigned simply because one small, otherwise inconsequential component was dropped from a particular manufacturer's portfolio.

Cost!

Some LEDs are expensive. Look for less expensive options: if there are none then consider whether your system can operate another way. An extra ten pence spent on a component can easily translate to an extra pound or two in the marketplace!

Construction

Construction is an important issue if you are to work with an optical system where a degree of imaging or beam control is required.

Most LEDs have an electrical contact attached to the centre of the emission area. With an imaging optical system this contact pad and its bonded wire are reimaged.

Rise time

For simple indicators the rise time, or speed of response, characteristic of an LED is not so important; however, if the LED is to be used within a communications system then the rise time will set a fundamental limit to the system bandwidth. Typically, rise times are between 15 and 500 ns for full output.

General remarks

To close this section, it should be noted that LEDs are difficult to control optically, because they are not a well-behaved source in terms of physical emission characteristic.

Most LEDs have a well-defined polar output, but on a diffuse background; so stray light may be a problem in applications that demand good rejection of background light levels.

The electrical contact pads to the LED can cause problems with imaging systems, as just mentioned above under 'Construction'. One simple way of controlling beam diameter and direction is to use a concentric series of apertures and strip away light where it is not required. This is not particularly efficient in terms of light utilization, but it is cheaper than a multi-element optical system.

Edge-light-emitting diodes

Most LEDs are surface-emitting types. However, a few are produced where the light emission occurs from the side of the device; these edge-light-emitting diodes (ELEDs) are particularly appropriate for the communications industry, where bonding to fibre optics or fibre ferrules is of interest.

The bonding, or pigtailing, of a fibre optic to the LED is particularly useful, because the communications worker has only to splice or connect the fibre to the system in hand. The alternative would be for the engineer to optimize the coupling efficiency of the fibre to the diode for every connection made.

The ELED used to be a popular device for communications, but it is now somewhat overshadowed by the superior performance of the laser LED. The manufactured volume of laser diodes is increasing annually, so their price is falling and hence the cost advantage of the ELED is being eroded.

The laser diode

Many of the attributes of the conventional light-emitting diode discussed above apply equally well to the laser diode.

Conventional LEDs emit light because of spontaneous emission at a semiconductor junction. Raising the current above a certain threshold value causes optical feedback to take over and stimulated emission to occur. Just as with the lasers discussed in previous sections, population inversion is required, together with some form of optical cavity within which to contain and control the laser emission.

The establishment of a population inversion requires a high drive current and is achieved through the liberation of more p- and n-carriers (holes and electrons). The laser cavity is established through the creation of reflective structures at each end of the cavity. One mirror is partially transmitting to allow the laser beam to escape, whereas the other is nearly perfectly reflecting. The little light that is allowed to leak through the near-perfect mirror is allowed to fall on a small photodiode. This photodiode can serve as a feedback element within a control loop to ensure, for example, a fixed light output. The high refractive index of the laser material itself will result in a relatively high Fresnel reflection from the cleaved ends of the device.

Before the onset of laser activity the diode operates as a rather inefficient LED. However, once the lasing threshold drive current is reached then stimulated emission occurs and laser output increases dramatically with drive current; see Figure 8.6.

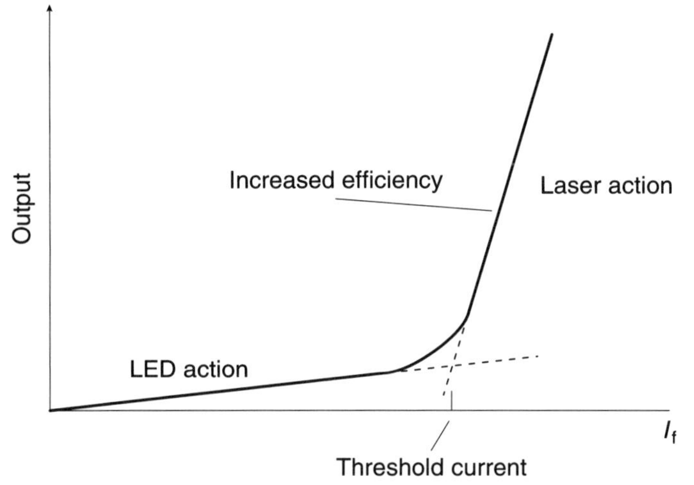

Figure 8.6 Laser threshold characteristic

Laser diodes fall to three main groups, shortwave, middlewave and longwave.

The shortwave devices operate at present in the region between approximately 620 and 980 nm, although the shorter-wave devices in this group operate only for short periods. Devices that operate at 635 nm are available commercially with acceptably long lifetimes, but with low power of the order 3 mW. However, the pricing of these devices is higher than their longwave and well-established colleagues. Development and market share and/or volume will of course reduce these prices.

Middlewave devices operate very approximately within the region 1100 to 1700 nm, and are used extensively for fibre optic communication purposes.

Longwave devices operate in the band between approximately 3300 and 30 000 nm. These devices require cryogenic cooling.

The actual devices are very small and are only cubes: literally a few tens of micrometres on a side. They dissipate a significant amount of heat, and if they are not bonded to a suitable heat sink they melt in a few hundreds of nanoseconds. The packaging of a laser diode usually involves bonding to a suitable heat sink within the package itself; care then has to be exercised by the user to ensure that the package itself can port heat efficiently.

The laser diode can achieve a very narrow bandwidth of emission, of the order 0.2 nm. This narrow bandwidth is only achievable at relatively high power, where a single mode can dominate; at lower drive currents several longitudinal modes can exist spread over an interval of several nanometres.

With high powers, thermal effects become important and the wavelength of operation can move to a higher value. Thus low drive current means bandwidths of several nanometres with low output, whereas high drive current gives a longwave-shifted, high-power, narrow-bandwidth output. Beware: semiconductor lasers can mode-hop with both temperature and drive current!

In common with the ELED semiconductor, lasers are excellent for bonding to fibre pigtails or even for mounting within a fibre connector, which further enhances usefulness within the communications industry.

The beam characteristic of a semiconductor laser is not nearly so well behaved as some of its counterparts, the helium–neon for example. The emission typically originates from a small rectangular region, perhaps as small as 1 μm by 5 μm or smaller still. The small size of this structure causes the emerging beam to be diffracted through a sizeable angle. The most diffraction is caused by the smallest dimension, and therefore a beam shape, roughly elliptical, as indicated by Figure 8.7.

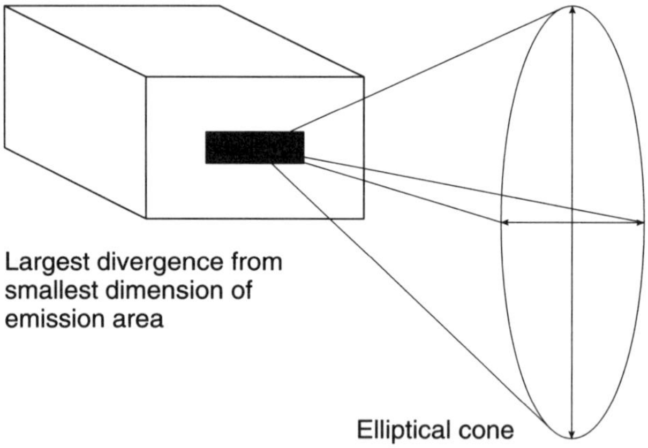

Figure 8.7 Semiconductor laser beam shape

Beam divergence is roughly in the ratio of 2 : 1 or 3 : 1, with extreme cases of up to 6 : 1. Beam half-angles from the small dimension of the emission area can be as little as 15 ° or may approach 90 °; the half-angle from the long dimension will be less by a factor of about 2 to 6 times.

To control the naturally diverging beam, some form of optical system is required. To control the asymmetric divergence, cylindrical lenses or prism systems can be used, (see Figure 8.8). With the prism system it is usual to collimate the light output first and then expand the minor axis of the ellipse to match that of the major. Simple but

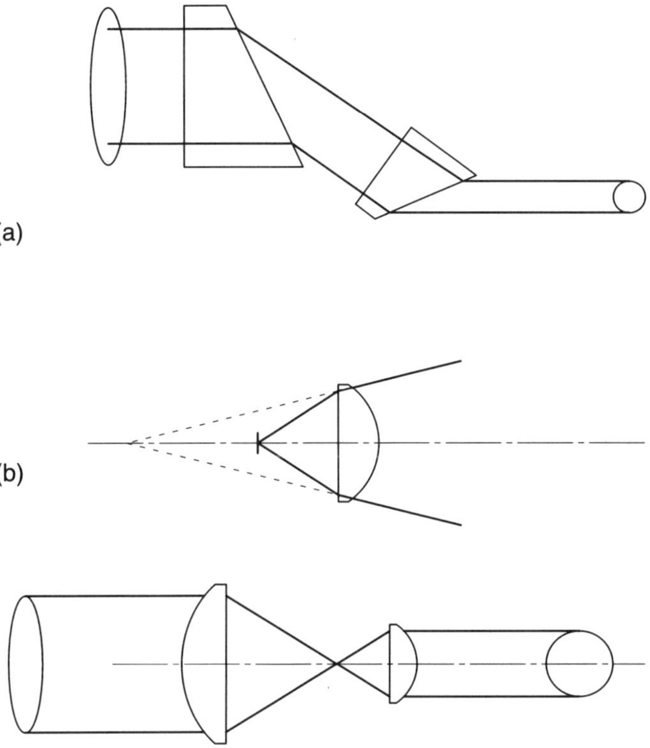

(a)

(b)

(c)

Figure 8.8 Beam circulization. (a) Prism pair (for use in collimated light): compression, or expansion, in one axis only. (b) Cylindrical lens: lens modifies the rapidly diverging beam. (C) Cylindrical lens pair: to circularize in collimated light

crude circularization can be achieved with a single positive cylindrical lens that slows the faster diverging beam. Alternatively, an anamorphic and afocal telescope can be constructed from two cylindrical lenses.

Some diodes exhibit a degree of astigmatism where light from the broad dimension of the stripe appears to originate from a small distance behind the exit face of the laser. The orthogonal light appears to originate from the surface itself. With non-index-guided laser diodes (beam containment within the cavity achieved by refractive index variation), the magnitude of the astigmatism could be as much as 50 μm, whereas the gain-guided type exhibits virtually zero astigmatism.

Laser diodes require only a very modest power input compared with their more conventional relations. A 10-mW diode may require only 150 mW of electrical power. This feature makes the laser diode considerably safer from an electrical point of view than its counterparts with high voltage and current supplies. However, the low-power electrical requirements, essentially those of TTL components, make the devices particularly susceptible to power variations. Power spikes can either catastrophically damage the device or cause damage to the optical facet through increased optical power density. The laser driver circuitry should ensure that no damaging power transients can be applied to the device.

The output from a single laser diode is limited by the physical volume of material and the requirement to dissipate heat as efficiently as possible. One way of overcoming these drawbacks and to increase power from what is essentially a single device is to form a monolithic array of laser diodes. Such devices might contain up to a hundred or more laser stripes. Powers of about 10 W continuous output are available.

Applications for semiconductor lasers are many and varied, from simple laser-pointing devices through to high-speed, high-reliability communications devices. High-power arrays can be used as laser pumps. Perhaps one of the most familiar applications is the data-read transducer in the compact-disc player.

Flourescent sources

The fluorescent tube is often overlooked as a scientific light source, but under certain circumstances can prove to be most useful. The types available range widely, from the increasingly popular compact tubes through to the more familiar room-lighting variants.

The fluorescent light source is a low-pressure discharge lamp with phosphors coated on the inside of the envelope that are activated by the discharge. Depending on the blend of the phosphor and the type of gas fill, various spectral outputs are available. When an electric current is passed through the gas an ultraviolet emission excites the phosphor, which in turn emits, usually, visible light. The UV emission is blocked by the glass wall of the fluorescent tube.

The fluorescent tube requires a ballast, which is a device for controlling the current flow through the tube. Usually the ballast will comprise an inductor together with the switch gear for actually turning the light source on or off.

Generally speaking, filament lamp output falls about 10 per cent over the life of the lamp. With a fluorescent lamp output is unstable over the first 100 hours, after which the 'initial' lumen value is measured. From 100 hours to 2000 hours light output falls exponentially to what is termed the lighting design lumen (LDL) value. After that, depending on the lamp type, depreciation is at between 2 and 6 per cent per 1000 hours of operation. A judgment has to made as to when the lamp has actually exceeded 'life'. This might be when light output has fallen to an unacceptably low value, when it is more economic to change the lamp rather than let it consume electricity, or when the lamp actually fails. See Figure 8.9.

A standard, warm white 40 W fluorescent tube will have an efficacy of the order 65 lm W^{-1}, with a flux of about 2500 lm (LDL).

There are many types of tube available, each with a different spectral characteristic. More recently the compact fluorescent type has been made available. These lamps, usually in clusters of two or four small tubes, have miniaturized electronic control gear with a high operating frequency of the order of 45 kHz as compared with the traditional tube's frequency of 50 Hz. Generally speaking, the new compact tubes give the same light output as their larger relations, but at 20 per cent of the electricity cost.

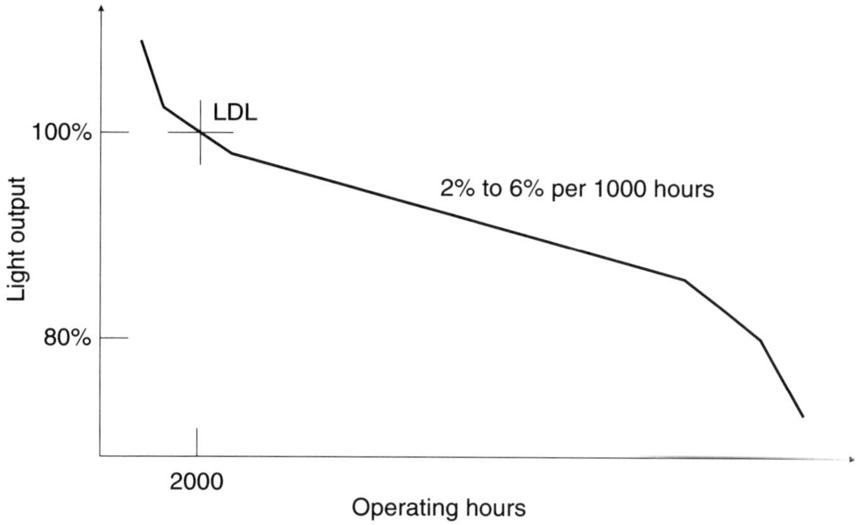

Figure 8.9 Lumen depreciation

This, together with a substantially increased life, makes the compact fluorescent an attractive commercial proposition.

Many types of tube are available, from black lights (UV emission only) through to types that model natural daylight with warm to cool variations in between. The spectrum emitted is somewhat spiky, owing to the emission characteristic of the particular phosphor coatings. A typical spectrum is illustrated in Figure 8.10.

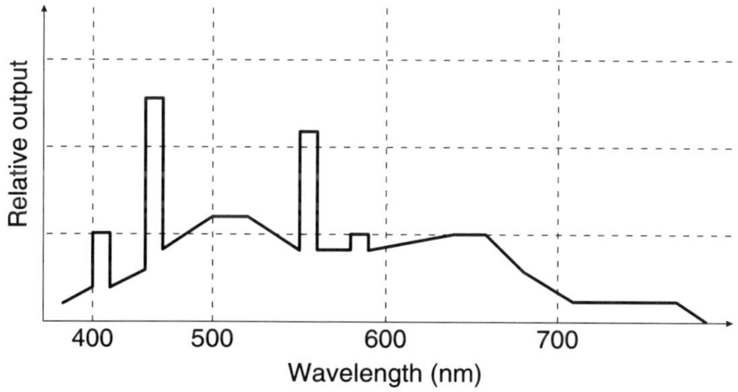

Figure 8.10 Generalized spectral characteristic fluorescent light source

Blackbody radiation

The most common type of light source is that described as a blackbody; such sources include incandescent lamps of all types, stars, and any matter that emits light when heated. By definition, a blackbody emits light in all directions and is therefore also a Lambertian source.

In fact, filament lamps are not true blackbodies but greybodies; their output spectrum is closest to a blackbody in the visible region, but they emit less infrared than does a true blackbody radiator. This deviation is not usually important, because most applications are for the visible spectrum, but the designer must be careful if a radiation source is required to emit light in the region above 5 μm.

A combination of Lambertian emission and a spectral output that can be described mathematically means that incandescent lamps are easily simulated in photometric models. One of the following conditions can usually be used to good effect:

1. Multiplying any blackbody spectrum by the photopic sensitivity curve ensures that the integrated output equates to the luminous power (in lumens) of the lamp.
2. If the output lumens are not known then the integrated output of the two curves can be used to determine the relative efficiency of an optical system containing components which have a transmission spectrum, i.e. the ratio of the initial luminous output of the lamp to the output of the system.
3. The total radiant exitance can be calculated and the luminous efficacy used to convert the watts per square metre into lumens per square metre. If the source area can be determined then the output power of the lamp can be calculated.

Blackbody sources are not inherently efficient as visible emitters; typically only 10 per cent of the total emitted light is in the visible spectrum. Electrically they produce only 17 to 35 lm/W, depending on gas fill, filament design and filament thickness.

Tungsten halogen lamps

The most common filament lamps are tungsten halogens. Tungsten melts at 3650 °C and so places an upper limit on the colour temperature that can be achieved with this type of lamp. The halogen gas has a beneficial effect on the lamp lifetime by recirculating evaporated tungsten back onto the filament; this is called the halogen cycle. The main causes of failure of incandescent lamps are either mechanical shock or else the occurrence of a 'hot spot' on the filament that causes it to melt. Not all parts of the filament run at the same temperature, but during nominal operation the difference is not significant. However, in some optical designs the lamp may be underrated, for example during a test with a prototype lens that cannot tolerate exposure to full power. In this case, great care must be exercised to ensure that the far field intensity pattern is representative of the lamp at full power. This point is particularly relevant when it is necessary to meet an even intensity distribution for products such as TV floodlights, landing lights or film projectors.

Here are some points to remember about incandescent lamps:

- Tungsten-halogen bulbs often have a quartz envelope that should not be handled directly; grease from the hand can cause the quartz to break when it is being used.
- Typically, colour temperature varies from 2700 to 3200 K; generally the higher the colour temperature, the thinner the filament and the shorter its lifetime.
- Very high colour temperature filaments, 3400 to 3600 K, can be obtained, but lifetime is of the order of 15 hours.
- Filaments are usually axial or transverse with respect to the centre of the bulb, but

the position of the filament within the glass envelope can vary by several millimetres. If alignment is critical then you should not rely on the nominal position of the filament.

- The output power of an incandescent lamp is proportional to the input voltage raised to the power of 3.5. This means that small voltage variations from the nominal operation conditions cause rapid changes in the luminous output.
- Underrunning the lamp voltage reduces both the luminous output and the colour temperature of the filament.
- Overrating the running voltage severely reduces lifetime.

The formula and constants that follow will allow you to design photometric models with confidence and to accurately predict signal levels or signal ratios for an optical system.

Spectral radiant exitance

The radiant exitance of a blackbody in $\mathrm{W\,m^{-2}\,m^{-1}}$ as a function of wavelength can be directly calculated using the Planck function:

$$M(\lambda) = 2\pi c^2 h/\lambda^5(e^{hc/\lambda kT} - 1) \tag{8.8}$$

In Chapter 6 we showed that $M(\lambda) = \pi L(\lambda)$, so the spectral radiance of a blackbody in $\mathrm{W\,m^{-2}\,sr^{-1}\,m^{-1}}$ is:

$$L(\lambda) = 2c^2 h/\lambda^5(e^{hc/\lambda kT} - 1) \tag{8.9}$$

The quantities, their values and common combinations are given in Table 8.6.

Table 8.6 Constants in blackbody radiation

Quantity	*Symbol*	*Value*
Velocity of light in a vacuum	c	$3 \times 10^8\ \mathrm{ms^{-1}}$
Wavelength in metres	λ	
Absolute temperature in kelvin	T	
Boltzmann constant	k	$1.380622 \times 10^{-23}\ \mathrm{J\,K^{-1}}$
Planck constant	h	$6.626196 \times 10^{-34}\ \mathrm{J\,s}$
Stefan–Boltzmann constant	σ	$5.66961 \times 10^{-8}\ \mathrm{W\,m^{-2}\,K^{-4}}$
First radiation constant (c_1)	$2\pi c^2 h$	$3.7418 \times 10^{-16}\ \mathrm{W\,m^2}$
	$2c^2 h$	$1.1911 \times 10^{-16}\ \mathrm{W\,m^2\,sr^{-1}}$
Second radiation constant (c_2)	ch/k	$0.014388\ \mathrm{m\,K}$
	$2h/c^2$	$1.4745 \times 10^{-50}\ \mathrm{W\,Hz^{-4}\,m^{-2}\,sr^{-1}}$

The Planck formula can be written in a number of ways, allowing the number of photons emitted to be calculated or substituting frequency (ν) for wavelength.

In photons $s^{-1} m^{-2} sr^{-1} m^{-1}$:

$$n_\lambda = 2c/\lambda^4(e^{hc/\lambda kT} - 1) \qquad (8.10)$$

In $W m^{-2} sr^{-1} Hz^{-1}$:

$$L_\nu = 2h\nu^3/c^2(e^{h\nu/kT} - 1) \qquad (8.11)$$

The formula can also be simplified for two special cases, as follows.

When $hc << \lambda kT$, the Rayleigh–Jeans approximation can be used:

$$L_\nu = c_1 c_2 T \nu^2 \lambda^{-4} \qquad (8.12)$$

The error in this expression is given by $1-[(e^{hc/\lambda kT}-1)\cdot \lambda kT/hc]$, and is less than 1 per cent provided $hc/\lambda kT < 0.02$.

Also, when $hc >> \lambda kT$, Wien's law can be used:

$$L_\nu = c_1 \lambda^{-5} e^{-hc/\lambda kT} \qquad (8.13)$$

the error in this equation is $-e^{-hc/\lambda kT}$ and is less than 1 per cent, provided that $hc/\lambda kT > 5$.

Stefan–Boltzmann equation

The total radiant exitance for a blackbody is calculated by integrating the Planck formula:

$$M_{\text{total}} = \int_0^\infty M(\lambda)\, d\lambda \qquad (8.14)$$

This integral is evaluated by defining the dummy variable $x = hc/\lambda kT$, with $dx = -d\lambda \cdot hc/kT\lambda^2$. The integral is now reduced to the form

$$\int_0^\infty \sigma T^4 \cdot dx/x^3 \cdot (e^x - 1) \qquad (8.15)$$

$(e^x - 1)^{-1}$ can be written as a series expansion and the result shown to be

$$M_{\text{total}} = \sigma T^4 \qquad (8.16)$$

This is the Stefan–Boltzmann equation, and σ, in $W m^{-2} K^{-4}$, is the Stefan–Boltzmann constant derived from

$$\sigma = 2\pi^5 k^4/15c^2h^3 \qquad (8.17)$$

Wien's displacement law

The Planck formula can also be used to show that a blackbody at a temperature T emits the maximum power per unit wavelength at a wavelength λ_m when

$$\lambda_m T = 2898\ \mu m\, K \qquad (8.18)$$

The position of the peak of the blackbody radiation curve is shown for a number of temperatures together with the form of the spectral radiant output in Figure 8.11.

Correlated colour temperature

Any light source whose colour coordinates match those of a perfect blackbody can be assigned an equivalent colour temperature. For example, a tungsten filament lamp might have a colour temperature of 2300 K, in which case its colour coordinates would be $x = 0.4957$, $y = 0.4152$.

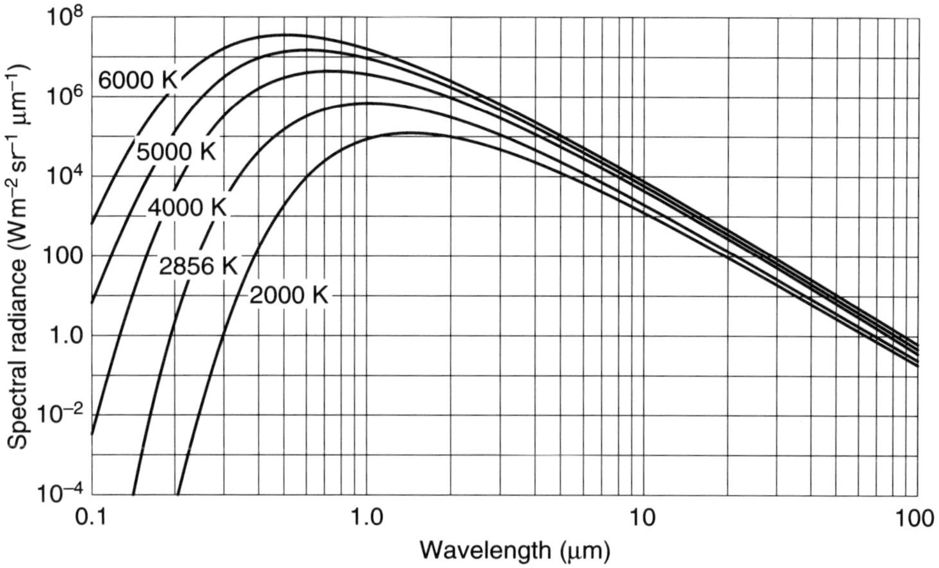

Figure 8.11 Spectral radiance of a blackbody at the absolute temperature shown

This method of assigning a colour temperature to a source cannot be used when the radiation does not follow the Planck equation. In this case a correlated colour temperature (see CIE Publication 17.4, listed in Further reading) is used to provide the closest match between the actual source colour and a blackbody.

Correlated colour temperature can be determined graphically by matching the chromaticity coordinates of the source to the nearest isothermal line on a chromaticity diagram such as Figure 6.8 or 6.9.

Alternatively, correlated colour temperature can be calculated by varying u_p and v_p until the following function is a minimum:

$$(u_s - u_p)^2 + (v_s - v_p)^2 \tag{8.19}$$

u_s and v_s are the CIE 1960 colour coordinates for the light source, while u_p and v_p are those of the equivalent blackbody. Minimizing this function is relatively simple with a PC, but the result is often no more useful than the graphical method.

High colour temperatures such as the northern sky (20 000 K) are predominantly blue. A good daylight white source such as a xenon arc lamp would have a colour temperature of between 5000 and 6000 K. Tungsten filaments vary from 2300 K to about 3200 K depending on power, gas fill and application.

Discharge lamps

Discharge lamps are the generic description for spectral lamps and all types of long and short arc lamps. The general design principle is to confine a base gas with dopant inside a glass or quartz envelope containing two electrodes, usually tungsten. Sometimes the gas is pressurized at the time of manufacture (xenon), or it may attain an overpressure during operation (mercury). An electric current is passed between the cathode and anode, exciting the gas into a plasma. The electric current can be either d.c., in which case a hot spot is created at the cathode, or a.c., which has two hot spots at either electrode.

Arc lamps are very important light sources because of the high brightness they produce and also for their precisely defined output spectrum, which is largely invariant with time and in the case of xenon almost constant through the visible spectrum. This attribute makes arc lamps ideal sources for photography, film and TV projection, colour LCD illumination, scientific instrumentation and many other optical systems.

Short arc lamps

Probably the most commonly used type of arc is the short or compact arc lamp, so called because the electrode separation is relatively short. The electrode separation depends on the base gas, electrode materials, voltage and input power, but is typically anywhere between 1 mm and 15 mm.

Often an average luminance is quoted for the arc lamp. In this case, the emission area is that defined by the distance between the electrodes during operation and the width of the arc at the point where it is half of the peak intensity. A mean reading of the luminance is taken over this area, and is usually quoted in candelas per square centimetre ($cd\,cm^{-2}$). Typical average luminance can vary from 25 000 to 220 000 cd.

Design hints and problems:

As with nearly every light source, increasing the nominal input power results in an increase in total output lumens, but also an increase in emission area. The net result is that the input electrical power may double but the brightness remains constant or even falls, and the output radiation from the optical system remains at best unchanged. Be warned: it is sometimes better to specify an arc with a lower power but higher luminance than to use a larger arc with more luminous power but a lower brightness. Usually d.c. arcs have a higher average luminance than their a.c. equivalent, and the absence of a second hot spot makes them easier to design into a condenser or collection optic.

Arc lamps emitting intense radiation in the UV can create ozone. This can present a serious health hazard, so the designer must take care to ensure that it is correctly dealt with or that an ozone-free lamp is specified.

The output power of all arc lamps reduces with time, largely because of electrode wear. Most manufacturers quote the initial luminous power from the lamp, so it is wise to allow a degradation factor of up to 80 per cent over the first 100 hours' use.

All arc lamps produce a high output in the near-infrared spectrum; for most optical designs this serves no useful purpose; rather it can heat up every component and melt plastics or even the epoxy holding adhesive bonding a fibre bundle together. Only heat-rejection filters, either as discrete components (hot/cold mirrors, heat-absorbing glass) or as coatings on some of the components, will remove the IR before it destroys a susceptible element. Inevitably, IR removal is accompanied by a loss of red light; this can change the colour temperature of the source and may lead to a loss of colour-rendering capability. Removal of heat is nearly always a compromise.

The two most common types of compact arc and their uses are now described.

Mercury arcs

Efficacy varies between 40 and 60 $\mathrm{lm\,W^{-1}}$, yielding 1300 to 30 000 lm from between 50 and 500 W input power, at a.c. and d.c. voltages. Average luminance is in the range 30 000 to 20 000 $\mathrm{cd\,cm^{-2}}$. These lamps are most often used in microlithography, where the high UV–blue output from the mercury lines is required. Other common uses include fluorescence assay illumination, microscopy and endoscopy.

Xenon arcs

Generally the efficacy is about 40 $\mathrm{lm\,W^{-1}}$, but some of the low-powered arcs may only produce from 18 to 20 $\mathrm{lm\,W^{-1}}$. Input powers range from 75 W to 10 kW, a.c. only, giving a luminous power range of between 1000 and 500 000 lm and average luminances of 15 to 100 $\mathrm{kcd\,cm^{-2}}$. The output of a xenon arc is continuous and nearly constant over the whole visible spectrum. Xenon has a correlated colour temperature of around 6000 K, which is close to Standard Illuminant D_{65} (see Appendix A) and makes xenon an ideal choice as the light source for film, slide, 8 mm, 16 mm, TV , video and LCD projection. It is also used in monochromators, spectrophotometers and solar simulation.

Metal halide lamps

These are high-pressure discharge lamps containing a base noble gas, mercury, and certain metal halides from the rare-earth group such as holmium and thulium. This combination produces a multiline wavelength distribution throughout the visible spectrum, resulting in a colour temperature of 5600 K. This makes this type of lamp ideal for applications in which artificial daylight is required such as theatre, film or TV lighting and photography. Metal halides are also used in some colour LCD projectors because of the relatively small amount of heat they produce. Lamp powers vary from 125 to over 12 000 W, and with the efficacy in the range 60 to 96 lm W the luminous powers available range from 10 000 lm to 750 000 lm. Arc length varies from 5 to 14 mm, and typically the average luminance of a metal halide is lower than that of the equivalent powered xenon.

The biggest drawback with metal halides is that colour variations can be seen in some lamp types when they are used as projection sources or spot lights. This is due to the differing weights of the various metal halides causing them to be located at physically separate positions in the plasma – orange to blue coloration is especially common. In spotlights the effect can be minimized by disrupting the surface of the projection lens or reflector containing the arc lamp. This is called ‘peening’ and is nearly always accompanied by a loss of system efficiency.

Long arc lamps

The most common types of long arc lamp are the krypton and xenon lamps. Usually water-cooled, the tubes typically vary in power from 2.5 to 5 kW, giving a broad spectral output from the UV to near IR. Intense spectral lines in the region 750 to 900 nm make krypton lamps ideal as a pump source for Nd–YAG lasers. These arcs are d.c.-operated, with emission occurring over the entire volume between the electrodes; typically they are 75 to 100 mm long by about 7 mm in diameter.

Spectral lamps

As their name suggests, spectral lamps produce line spectra defined by the inert gases and metal vapours which they contain. The base gas, such as argon or mercury, can be loaded with a metal such as cadmium, cesium, sodium, potassium, etc. to give a narrow-band line spectrum anywhere from 250 to 800 nm. The emission area is usually larger than in short arc lamps, typically 15 to 20 mm long by 6 to 8 mm wide, and consequently the brightness is not as intense.

Spectral lamps are usually used with narrow-band filters to produce monochromatic light sources for applications such as spectroscopy, fluorescence assays, and medical and chemical analysis.

CHAPTER 9

Mechanics

Introduction

Mechanical mounting techniques warrant much more space than we can offer in this book. Consequently, the aim of this chapter is to present some of the main points that must be considered during the mechanical design of optical components and to explain the relevance of some of the common terminology.

Specifically, we address the mechanics of holding a lens, or lens system, in place. This is of course not just for the laboratory but for the real world where vibration, shock and thermal effects will conspire to disturb your carefully mounted lenses.

It is one thing to prove the performance of an optical system in the laboratory where all degrees of freedom are available and the temperature is constant, but quite another to take that arrangement and mount it in an economically manufactured housing. It is in this area that the optical designer has to work very closely with the mechanical designer.

After a lens design has been roughed out, the optical designer should conduct a high–level tolerancing exercise to identify if any of the components are particularly sensitive to position, tilt or rotation errors. Any component so identified should be discussed with the mechanical designer so that a decision can be made as to whether the design is going to be practical to manufacture. If it is not then the optical design must be changed or a novel mechanical system identified. This principle applies equally well to the assembly of the lens system.

Towards the end of the process of designing a lens, and as part of the tool fitting, a final tolerancing exercise will be conducted. This will have the aim of providing a table of positional and rotational error budgets to the mechanical designer. In this way the required mechanical quality of the system can be identified. At one end of the quality spectrum will be a simple, plastic, push-together mounting system; at the other will be complex mounts that allow adjustment of the optical elements during assembly. This latter option will almost certainly include dynamic testing of the optics during assembly to ensure that the required performance is achieved.

For any worker contemplating the manufacture of optical systems on a production basis, mechanics is a vitally important area. The mounting must be kept simple and cost-effective to maintain profitability. However, if the lens assembly doesn't work or it falls to pieces during use then the product is not going to sell.

Lens drawings and specifications

Before we discuss the salient features of good mechanical design we should look at how the optical element is specified. In the UK the British Standard for this is

BS 4301 (1982), which covers the detailed specification of a multitude of optical characteristics. This standard will be replaced in the near future by European Standard ISO 10110. At the time of writing, approval has already be granted for much of the draft document and it seems destined to be adopted as a global standard. However, the basis of both standards remain common, and an understanding of them will serve the designer well.

Subject headings covered by all of these standards include

1. Form errors.
2. Centring errors.
3. Surface defects.
4. Surface texture.
5. Surface treatment.

To these we may add, when considering more than a single lens

6. Joining of elements.

A description of each item and how these specifications might be measured in practice follows.

Form errors

Lens form refers to the lens shape. Any error in form relates to the departure of the actual shape achieved in manufacture compared with the ideal shape as identified by the optical designer.

Form errors can arise from several different causes – worn tools, worn machinery, overpolishing, or even human mistakes. The tolerancing exercise conducted during the optical design will have identified a tolerance within which the form of the lens surface must lie.

The simplest form error is the tolerance on the radius of a spherical surface, which is usually presented in such form as

$$R\,50.0 \pm 0.1$$

This means that the radius is 50 mm, with an allowed tolerance of 0.1 mm. The mechanical drawing might look like that given in Figure 9.1. Lens dimensions are usually given simply as numbers, with a label somewhere on the drawing stating that they are all in millimetres or all in inches, rather than the units being written out each time.

Form error of both spheric and plane surfaces will be stated in terms of fringes, or as a number of half-wavelengths, and will be given on the lens drawing in one of two ways (see Figure 9.1; the symbol of the first method and number 3 of the second simply mean 'allowable form error').

How do we measure the spheric curvature or flatness of our planoconvex lens? There are a few techniques we can employ, three of which are fairly common:

1. Test plate.
2. Interferometric.
3. Newton's spherometer.

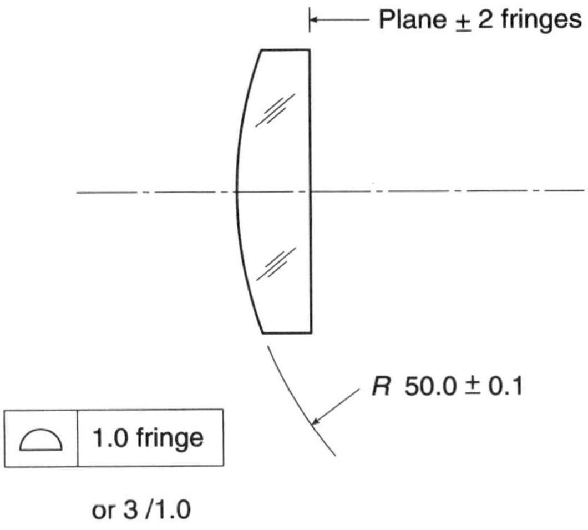

Figure 9.1 Specification of radius and form errors for a spherical surface

The test plate method uses the principle of Newton rings to find out if the form of a concave or convex surface is within tolerance. This technique can be used on or off the polishing machine. Newton's rings are seen when two optical surfaces of similar, or near-identical, form are brought in close contact and illuminated with monochromatic light. With reference to Figure 9.2(a), the rings originate from interference of the light reflected from the lower surface of the test plate and from the upper surface of the component under test. The size of the air gap decides the magnitude of path difference and hence whether constructive or destructive interference is observed. For a perfect component the air gap will not exist and no fringes will be visible. The transition between two dark fringes represents a change in path of one wavelength or a change in distance between surfaces of 0.5 wavelength due to the double pass across the air gap.

The test plate is a very accurate optical surface of the form that you are trying to manufacture. Master test plates are held for nearly all of the radii on a manufacturer's tool list.

An arrangement commonly used for the observation of Newton's rings is illustrated in Figure 9.2(b). The fringes are localized in the region where test plate and component are in contact. All the observer has to do is look at this position, recognize the fringes and count them. When the surface under test matches, or nearly matches, the surface of the test plate then the number of fringes observed is minimized and a decision can be taken on whether the surface form is acceptable or not.

The second technique, that of interferometry, is again a weighty subject of some substance. However, in very simple terms, as with the method of Newton's rings, interference fringes are produced that can be interpreted to give quantitative information with respect to the surface form. A full interferometer, such as a Twyman–Green or a Fizeau type, is a powerful tool for debugging complete optical systems as well as individual components. Several industrial versions exist with camera–computer interfaces for automatic assessment of the fringes so produced.

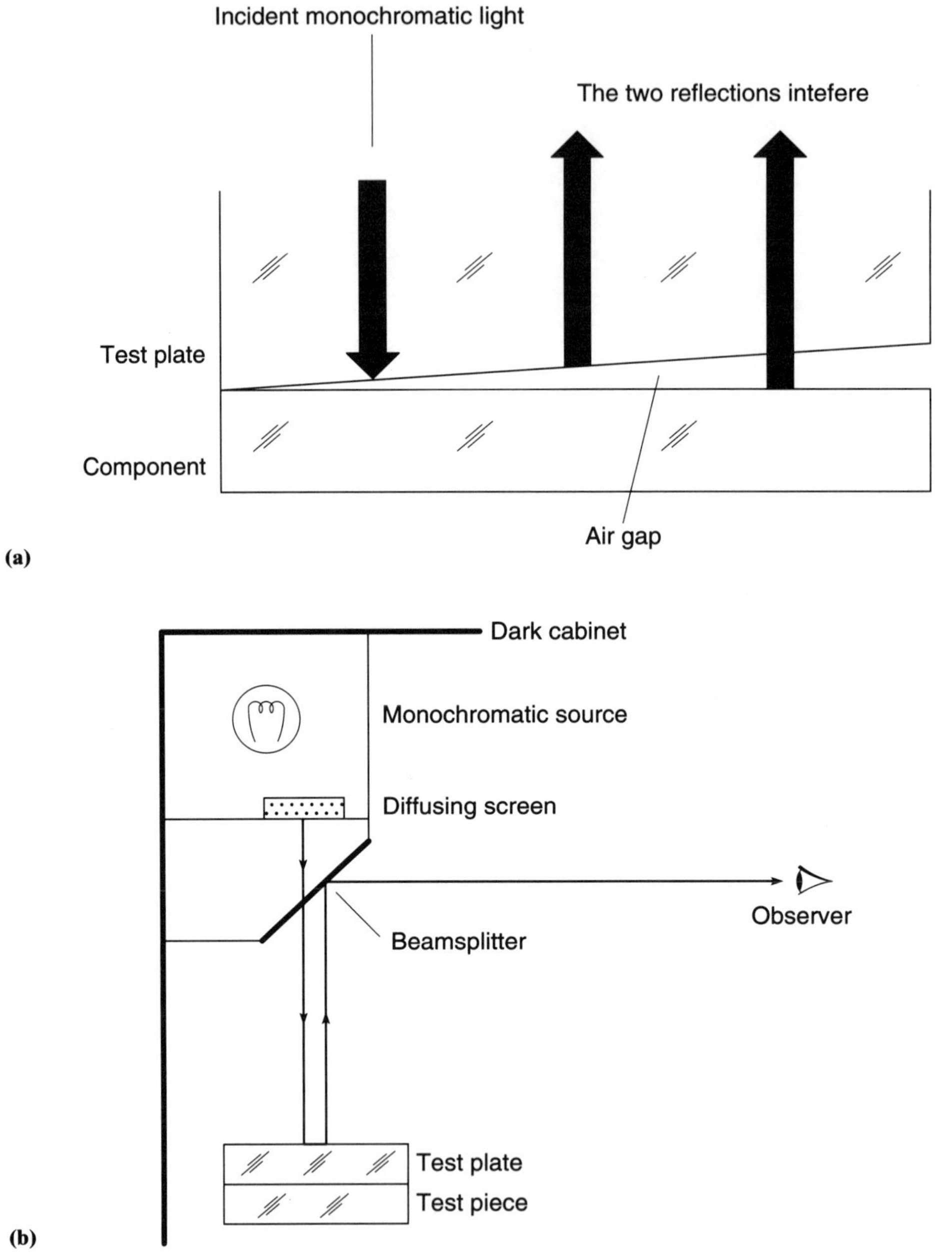

Figure 9.2 (a) Generation of Newton's rings. (b) Observation of Newton's rings

Finally, a mechanical technique can be applied to measure directly the radius of a spherical surface. A tripod arrangement known as a spherometer is placed on the surface of interest and a micrometer adjusted, or a plunger depressed, to measure sagittal height or the power in diopters (see Figure 9.3(a) and (b)). The circular ring defined by the feet of the tripod is known and fixed, and through calibration the

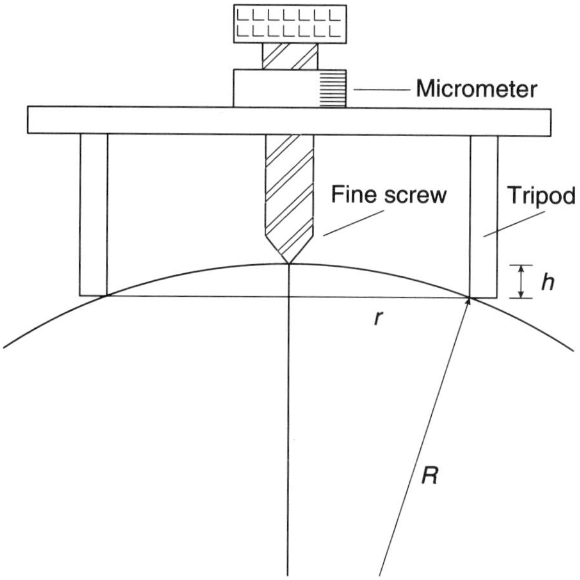

(a)

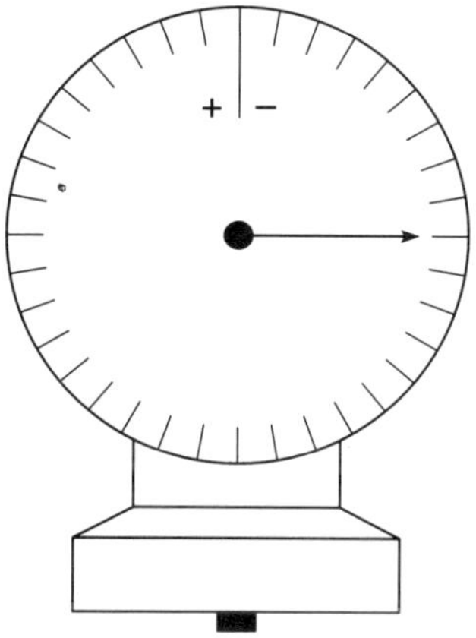

(b)

Figure 9.3 (a) Newton's spherometer (b) Spherometer: plunger type

sagittal height is measured. The following relationship can then be used to find the radius:

$$h = R - (R^2 - r^2)^{1/2} \tag{9.1}$$

where h = sagittal height
R = radius of curvature of surface
r = radius of spherometer ring.

Alternatively, with the dioptre ring, once the power has been measured and you have a dioptre measurement, the following relationship can be used to find the radius. (Be careful: the dioptre gauge (plunger type) is not a very accurate instrument and the measurement tolerance is often no better than ±0.1 at best.)

$$R = [1000(n-1)]/D \tag{9.2}$$

where R = the radius of curvature
n = the refractive index of the component
D = the gauge reading.

(Remember, the dioptre is the reciprocal of the focal length in metres, hence the factor of 1000.)

Thus if a surface power of two dioptres is measured and we take $1000(n-1)$ as 525 for an average ophthalmic glass the radius is

$$525/2 = 262.5\,\text{mm}$$

The key feature for mechanical designers will be converting the tolerance on the radius into a maximum and minimum variation at the point of contact with the lens. For example, if a lens is held against an internal radius, then its effective position changes with any variation on the radius of the contacting face. This in turn can cause a change in the air gap between lenses and a tolerance build up that can result in both optical and mechanical problems. Check the effect of radii tolerances by using the spherometer formula in Equation (9.1), to calculate the sag of the surface.

Centring errors

A centring error occurs when the axis of the optical component is not coincident with the optical axis of the system (see Figure 9.4(a)) or when the two surfaces of the same optical element are not concentric (this is the same as tilting a spherical surface; Figure 9.4(b)).

It is impossible to reference centring tolerances to the optical axis itself and so a different, well-defined axis is established as the datum.

The lens concentricity (optical to mechanical axes) is most easily defined and measured by considering the deviation of a ray directed along the mechanical axis of the lens, then measuring the angle of deviation of that ray (see Figure 9.5). When the lens is rotated about its mechanical axis the deviated ray of light will describe a circle, the radius of which subtends an angle at the lens; it is this angle that is specified for the definition and measurement of centration. Many simple centration jigs exist for looking at the deviation of, say, an He–Ne laser beam directed at a screen placed at the focal distance of the lens.

In design drawings, the allowable centration error will usually be presented in two ways: surface-to-surface centration (or tilt) and mechanical-to-optical axis centration. The first is illustrated in Figure 9.6, where A and B define the datum axis; the symbol indicates allowable surface tilt and the magnitude is the allowable departure from squareness of that surface measured in minutes of arc. The second is illustrated in Figure 9.7, where the prefix '4/' means permissible lens centring error and '2' the error in minutes measured as described above.

From the aspect of mechanical design, lens centring is often accomplished by forcing

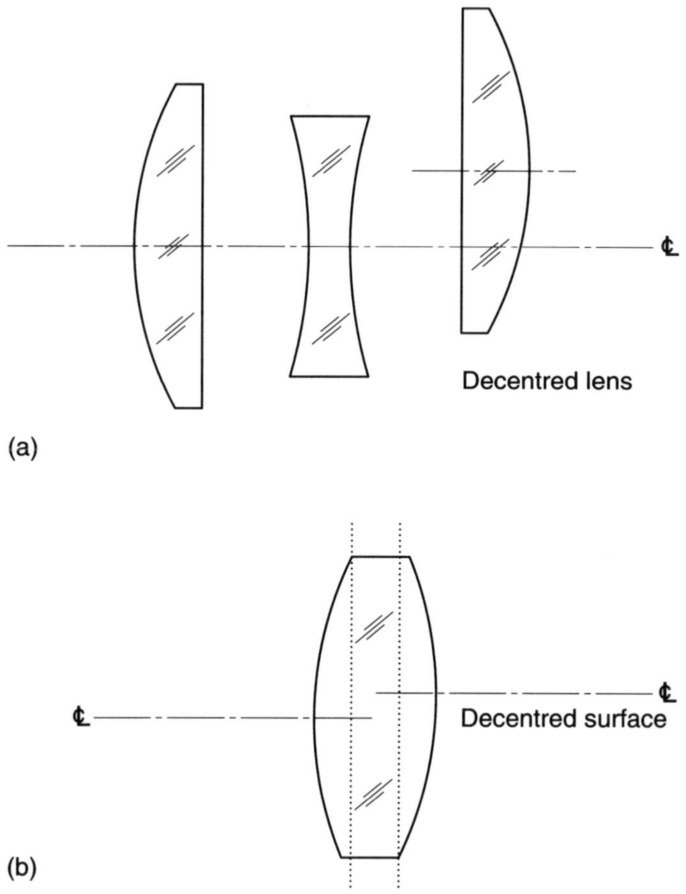

Figure 9.4 Centring errors: (a) decentred lens, (b) decentred surface

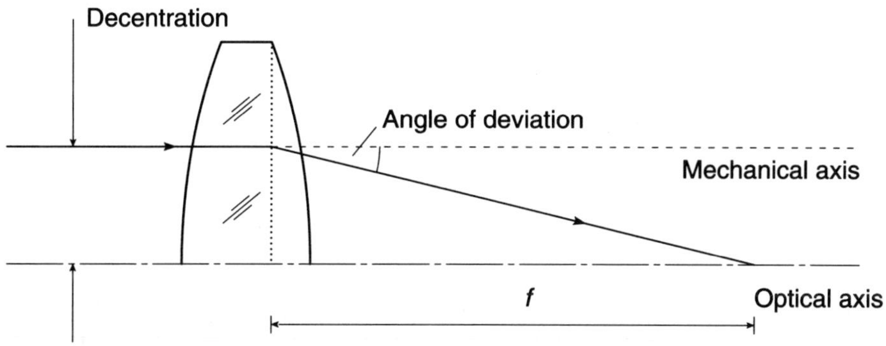

Figure 9.5 Measurement of centration error

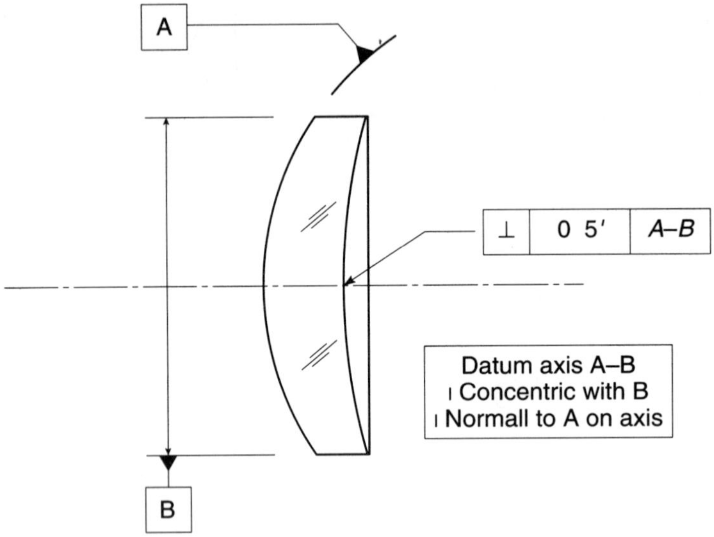

Figure 9.6 Surface tilt

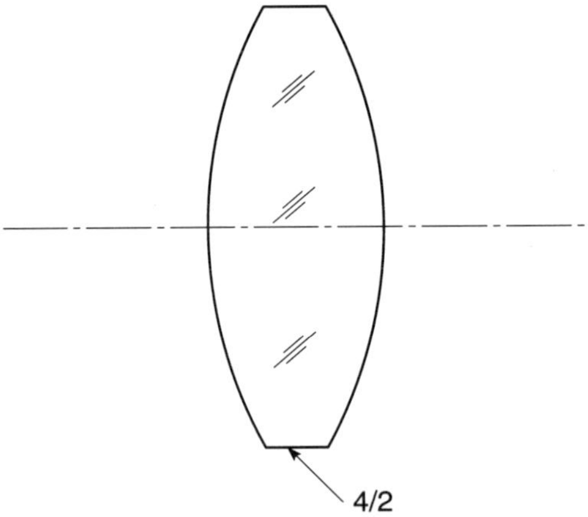

Figure 9.7 Component mechanical to optic axis centration

a spherical surface to contact against a bore that is slightly smaller than the maximum diameter of the lens. The lens is then forced against this internal diameter. Provided the radius of the lens is not very long, this method is effectively self-centring and works well if tolerances are not very tight. For high-precision centring, the lens can be actively aligned by hand and then potted into position or adjustment built into the design. The most common example of this method is to use three flat-ended grubscrews spaced around the diameter at 120° intervals. The flat ends of the screw bear against the edge of the lens and centring accuracies of about a minute of arc are possible.

Surface defects

Surface defects are usually described in terms of scratch and dig and refer to the surface quality of an optical component.

The definition and use of the scratch-and-dig defect rating codes have already been presented in Chapter 4 in the discussion of optical mirrors; the salient points may be recapped as follows.

The surface finish is usually described by two numbers, which refer respectively to scratch and dig, for example 60/40; they are derived thus:

- *First number, scratch.* This indicates the apparent width of a scratch. It is specified against standard scratches, such as a No. 80 standard scratch.
- *Second number, dig.* This indicates the apparent diameter of a dig, bubble, inclusion or similar flaw. A number, say 50, indicates a defect diameter of 0.5 mm, 20 a diameter of 0.2 mm and so on. Irregular digs are sometimes quantified by the sum of half the length plus the width.

Scratch-and-dig numbers form a surface defect code that is readily interpreted by all optical workers. The code is usually accompanied by an indication of the allowed number of defects on or in the component. Thus an 80–50 surface would be fairly easy to fabricate and so relatively inexpensive. A 20–10 or 10–5 surface is much more difficult to fabricate and limited to demanding applications, such as laser optics or interferometry.

Inspection of components for scratch and dig is always done visually against known reference standards. Defects are viewed in bright light against a dark background, and are directly compared against standard defects. It must be stressed that surface defect counting is a highly subjective activity and subject to interpretation. If you are having a coating applied to your optical elements then scratch and dig should be assessed both *before* and *after* coating.

Surface texture

Surface texture really falls into two different types of specification, one for optically grey surfaces such as diffusers and the other for optically polished, near-perfect surfaces. The two conditions are very different.

Grey surfaces are defined in terms of R_a value. (This has already been covered in Chapter 4 with respect to mirror surfaces.) The R_a concept, or centreline average method, is briefly recapped here.

The R_a value is the average height from a mean line of all heights of the surface, whatever the sign of the height.

$$R_a = \frac{h_1 + h_2 + h_3 + \cdots + h_n}{n}$$

$$= \sum h/n$$

It is usual to specify the area over which the flatness or roughness specifications are to apply, such as the clear aperture of the lens.

BS 4301 uses a tick symbol with which to denote the surface to be worked (see

Figure 9.8). 'N' numbers above the tick refer to a grey surface, whereas a number within a kite symbol refers to the grade of polish. The numbers are given in Table 9.1.

Table 9.1 Roughness numbers and polish numbers

Roughness number	R_a *value* *µm*	*Polish number*	*Grade*
N3	0.1	1	Coarse
N6	0.8	2	Medium
N9	6.3	3	Fine
N12	50.0	4	Very Fine

Care must be exercised in specifying surface finishes; a surface with, for example, scratch and dig of 10/5 and to polish '4' standard will be difficult and time-consuming to fabricate. This sort of specification would be applied only to the very highest-quality optics, for example those used as very high-quality reticules. The designer has to decide what is the minimum specification that will allow the optical system to be fit for purpose.

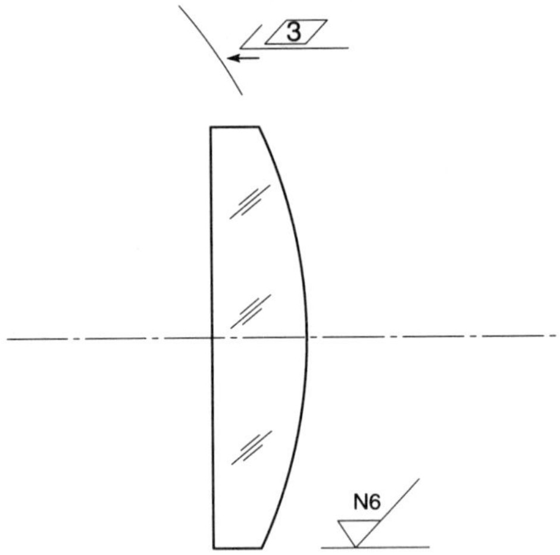

Figure 9.8 Surface finish notation

Surface treatment

Surface treatment refers to coatings, which come in three types: reflective, antireflective and protective.

A reflective coating can be specified as 'first' surface, 'second' surface or as a partial coating. The symbols commonly associated with each type are shown in Figure 9.9. With an antireflection coating it is usual to specify the centre wavelength for minimum reflection; if this is not stated then 546 nm is usually assumed. If the component is to operate with a non-normal angle of incidence then this should be indicated on the drawing (see Figure 9.10). Other finishes, including blackened, painted or protective surfaces, are usually described with arrows and notes.

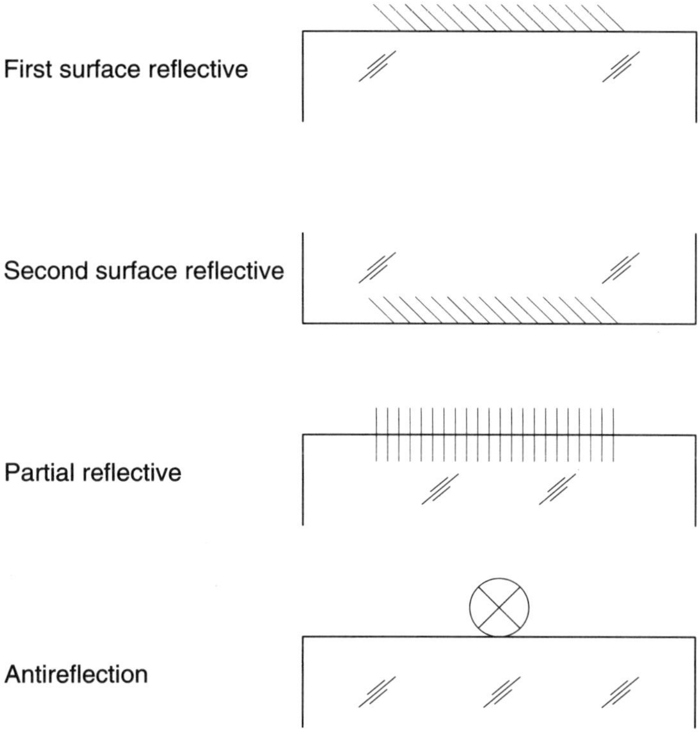

Figure 9.9 Coating symbols

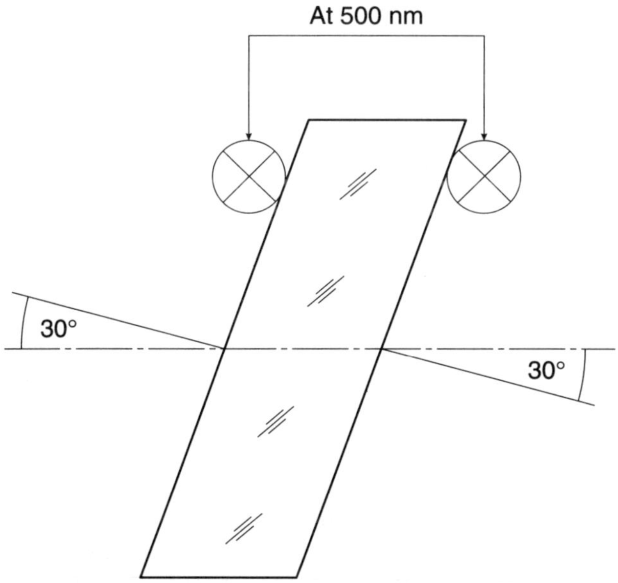

Figure 9.10 Coating specification; this example shows the specification of an antireflection coating for an angle of incidence of 30° and a wavelength of 500 nm

Joining of elements

When elements are joined, such as a cemented achromatic doublet, the quality and cement instructions have to be presented on the drawing. Each element and the type of cement to be used should be indicated.

The cementing process itself will introduce blemishes and other defects. These should be covered by accompanying notes and specifications as described above.

There are many different optical cements available, with a large range of refractive index and temperature capabilities. It is best to take advice from the lens manufacturer about which to use according to your circumstances. However, it is essential that the optical cement selected has a thermal expansion coefficient similar to those of the glasses it is cementing together – otherwise you may end up with two single lenses again.

This section has dealt with some aspects of specifying and testing optical surfaces. The discussion has been limited to consideration of spherical lenses only and has not looked at the testing of aspheric optics. Aspheres demand a different testing regime based on the same tools and techniques but with a different interpretation of the results obtained. For example, by definition the aspheric surface is not spherical and hence a test of Newton's rings type will show departure from the ideal sphere. For aspheres it is a matter of 'Is it the right amount of departure?' rather than one of minimum departure.

Lens mounting

Mechanical design for lens mounts usually falls into one of two distinct areas. The first is for laboratory usage, where a uniform temperature and optical friendly environment should exist, although this is not always true. The second reflects the harsh realities of the real world. A lens used in the laboratory environment will usually not have to cope with temperature, humidity, shock, abrasion or vibration, whereas a production window for a supersonic aircraft will. The mechanical mount has of course to withstand the same environment and to a large degree protect the optical element itself.

Simple supportive and clamping structures for use in the laboratory can be purchased from commercial component suppliers. Each supplier has its own kit of parts and its own way of doing things. However, there does come a point when an optical rail or bench cannot do the job that you require it to do. So you might have the job of mounting a mirror for an interferometer or a lens to get into an awkward space.

One mounting technique to avoid is that of double-sided sticky tape. One of the authors once mounted a mirror for an interferometer application by using double-sided tape between the mirror and a plate mounted on a vibration isolating pole. The mounting technique worked well throughout the duration of the experiment, but there was no way in which the mirror could be removed from the plate! Final resort was to use an ultrasonic bath together with a solution to dissolve the adhesive. It took the best part of a week to remove the mirror! In summary: take care with this type of tape and only use the minimum necessary.

A mirror mount for the laboratory simply has to support the mirror, perhaps on a block, and then have some means to prevent the mirror falling off the block (see Figure 9.11). Many variations exist on this theme, including retaining rings that fit all the way round the mirror as a collar. The basic rule is the same: support and retain. For interferometric purposes, one key rule to remember is to avoid distorting the mirror as this will affect the interferometer's performance. It will take very little force to distort an interferometric mirror. The mounting plate shown in Figure 9.11 can of course be modified to fit to an optical rail, a table or any other form of bracketry.

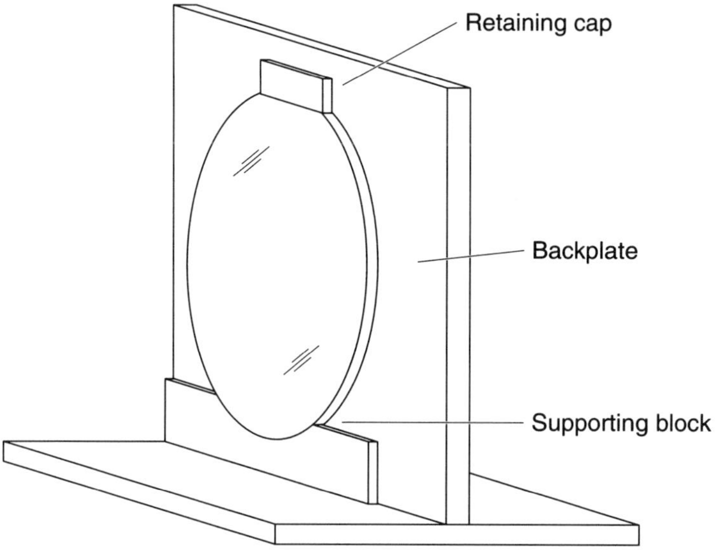

Figure 9.11 Simple mirror mount

Mirror mounts should have the minimum number of adjustments possible. The more degrees of freedom there are, the more there is to go out of adjustment. Kinematic mounting techniques can be applied here, so that mirrors can be demounted and then returned to position with the minimum of adjustment. Kinematic mounts are three-point mounting schemes, usually with one fixed and two adjustable points of contact.

For lenses a pocket and retaining ring works quite well (see Figure 9.12). Such a simple mount can be pin-mounted or attached to an optical rail. Here the lens sits in a pocket against a retaining flange. The clamping or locking ring slides in the bore to press against the lens. The whole is locked together through use of a few screws.

If a series of lenses are to be fitted together within a barrel arrangement then it might be possible to arrange for all of the turning and machining operations to be completed without removing the tube or bar from the machine. This will help with maintaining a good concentricity of components and lenses.

If the optical components are to experience a wide range of temperatures – for example, a hot light source is to be focused on or near a lens – then the component

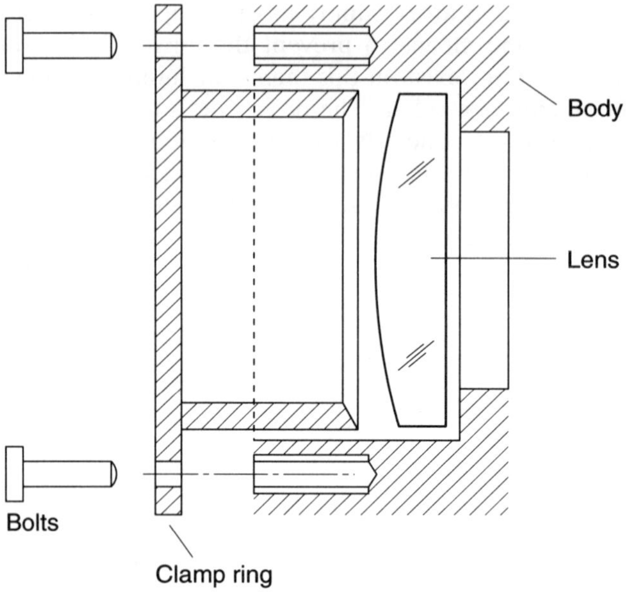

Figure 9.12 Simple lens mount

should be mounted loosely to give the glass, or whatever, room to expand on heating. If the thermal shock is large then it is good practice to ensure that the beam uniformly fills the component. A focus will result in an unacceptable degree of local heating and the component will break. Keep the heat even!

For commercial applications the mounting mechanisms require more sophisticated versions of the simple mount described above. Here the lens might be retained by a threaded locking ring (Figure 9.13), or cemented with a trough for cement spillage as in Figure 9.14. The cement for gluing a lens in place should be selected with care according to the range of temperature and humidity expected, because the glue must cope with any differential expansion between the mount and the lens itself. Care must be taken to prevent the adhesive coating the lens during assembly. If the lens is plastic then the use of 'superglues' (cyanoacrylates) should be avoided, because this type of adhesive releases gases over several days, which can cause the plastic to cloud over and become opaque; instead, use an acrylic dissolved in a suitable solvent such as chloroform, but be sure to follow the appropriate health and safety regulations and don't breath in!

These general mounting ideas can be extended to encompass, say, the mounting of several different elements or groups within the same concentric barrel.

Component mounting: fibres and prisms

Optical systems are not just comprised of lenses; many other components can be used. These may comprise prisms, plate beamsplitters, reticules, filters or polarizers, gratings, both reflection and transmission types, and fibre optics.

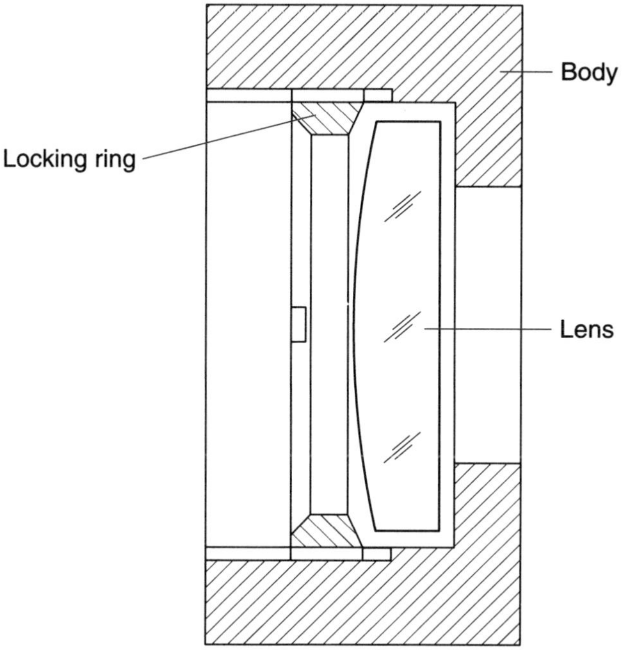

Figure 9.13 Threaded locking ring

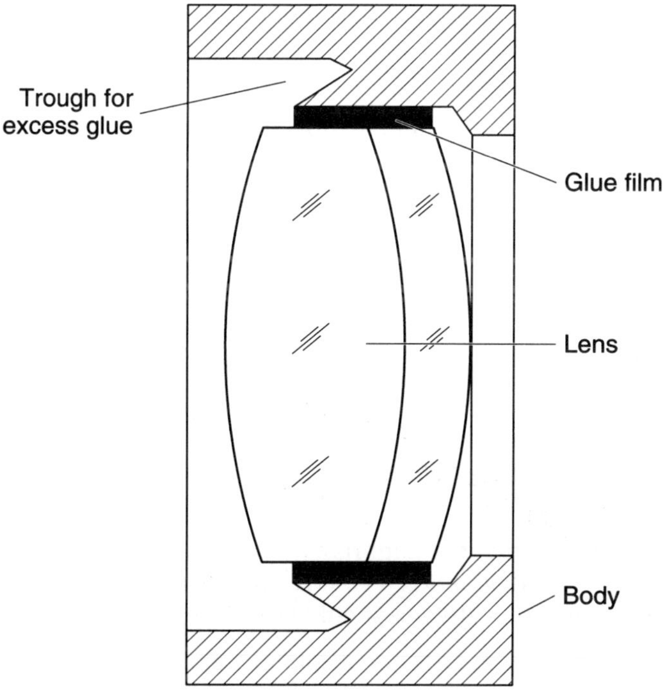

Figure 9.14 Cemented lens mount

In many respects the plano-optical elements can be mounted like lenses. However, remember that if a beamsplitter is mounted at 45° to the optical axis then the aperture will be elliptical and the mechanical mount must allow passage of the extreme rays, or vignetting will occur.

The general rule of 'support and retain' rather than 'clamp' applies.

Fibre optics

Fibre optics and prisms can be difficult components to mount so the following advice may prove useful.

The key to mounting fibre optics, either in a bundle or as a single fibre, is to do so without introducing stress or strain.

Single fibres can be potted in epoxy or mounted within ferrules – sometimes hypodermics are used – and then clamped. Note that it is the ferrule that is clamped, not the fibre, so there is no stress on the glass.

Some fibres, particularly those used for communications systems and some military applications, come complete with armour and steel support members. Not many optical workers will actually come into close contact with such fibres, but these well-tried and proven principles can be applied to lesser applications. If a small bundle of fibres is to be used then perhaps a steel wire within the same bundle can be used to provide strain relief for the bundle. An outer, armoured sheath can be used to the same good effect. These techniques apply equally well to coherent and light-guiding bundles.

All fibre optic cabling should be free from tight radii; the fibre specification should indicate a minimum bend radius through which the fibre can be bent without fear of damage or excessive losses.

For fibre optic bundles used for illumination purposes, it is possible that the fibre ends will experience high temperatures. Any cements used in this region must of course be able to withstand the same high temperatures.

Prisms

Usually prisms are relatively large and robust components but the same mechanical mounting rules apply: support, retain and protect. Two main mechanisms apply to prisms, i.e. clamp or glue.

Many clamping mechanisms can be envisaged from complex interlocking pockets through to the simple G-clamp. The G-clamp will be familiar to many from experience of spectroscopic experiments where the dispersing prism is retained on its rotation stage by a such a clamp.

With any clamping technique it is important to use only the amount of force that is necessary to do the job and no more. The forces applied should be distributed evenly across flat surfaces and corner stresses should be avoided.

Where precision is required, in terms of absolute angle or position, the prism must be mounted accurately within a regime which itself is capable of adjustment. One useful technique is to have a 'V' or right-angle block manufactured very accurately and glue the prism on this support structure. The structure should be castellated so that the prism can rest on the precision-ground parts with any excess glue allowed to run into the troughs (see Figure 9.15). The mount should be marginally larger than the prism so as to afford some mechanical protection against knocks and bumps.

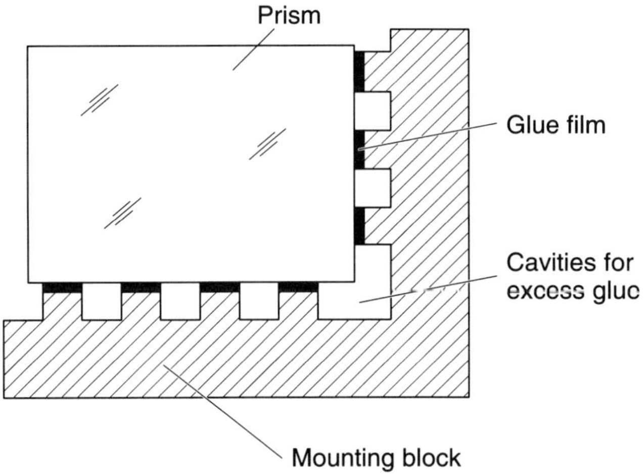

Figure 9.15 Section through castellated prism mount. In practice the castellations would be much smaller.

Any cements selected for gluing the prism block to its mount should be chosen with knowledge of the application borne in mind. This sounds obvious, but the glue may have to cope with temperature extremes, relative thermal expansion, removal in case of error, shock, high levels of ultraviolet and ageing.

For temporary mounting of prism components, wax may be used. It has the advantage of easy application, adjustment and removal with the use of a little heat.

Good practice

As always, common-sense rules apply.

The mechanical mounting system has not only to hold the lenses in place over a range of sometimes extreme conditions but also to control stray light, reflections and thermal distortion.

Light reaching the detector, or image plane, of an optical system that does not follow the intended design path is termed stray light. It arises mainly from reflections, especially at mechanical interfaces or refractive index boundaries. It can be thought of as a d.c. contribution to the signal level that causes a reduction in signal-to-noise performance or image contrast. While these effects may not be eliminated, it is possible to minimize them using one or more of the following techniques:

The first rule of stray light control is:

Work backwards from the detector or image plane

Use a dark finish, such as an electro-blacking or photographic blacking to reduce flare and stray light. A black anodizing, deep-black dye or electrolytic colouring process can be used to create a matt surface with a low reflectivity. Surface finish is just as important as colouration, and more so for grazing incidence applications.

Light at grazing incidence will reflect with little loss of power, so a rough finish or, better still, a screw thread cut inside the barrel will help to minimize stray reflections. In extreme cases a black felt is ideal for controlling unwanted reflections from mechanical parts.

The judicious use of baffles, apertures and stops will further reduce stray light. The application of antireflection coatings to all optical surfaces will drastically reduce stray light and ghost imaging.

For some applications that involve the use of a beamsplitter there may be a requirement for an optical beam dump. This is a device that allows light to enter but not to escape. Such a component is illustrated at Figure 9.16. The exact method of construction will depend on the application, but the basics of matt rough walls and an optical absorber, such as black felt, remain common to them all. We have even seen optical beam dumps made from a stack of razor blades laid one on top of the other; the cutting edges are directed towards the light source, so that multiple reflections between blades rapidly reduce the intensity.

Thermal tolerancing and the design of a suitable mechanical mount can be a fascinating subject. As temperature increases the refractive index of some optical

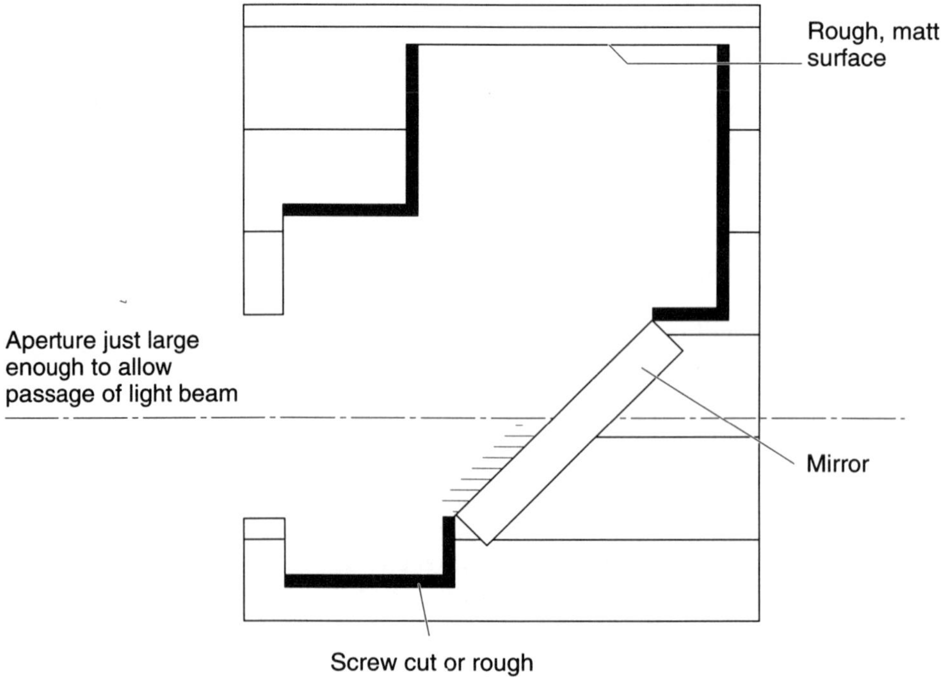

Figure 9.16 Optical beam dump

materials changes very rapidly so that the athermalization (invariance with temperature) of, typically, plastic lenses and those used in the infrared becomes important.

The change of refractive index with temperature for plastics is large and negative and of the order of $-110 \times 10^{-6}\,K^{-1}$. A reduction in refractive index means a reduction in the power of the optical element and a shortening of the focal length. Now, if the mechanical mount were to shorten as the temperature increased, then a self-compensating system could be established. This sounds simple but it is not. However, the technique has been used to good effect in certain infrared systems where the exotic materials used show similar index-versus-temperature characteristics. Alternatively, materials that reverse index change with temperature (positive and negative coefficients) have to be sought so that thermal correction can be achieved in much the same manner as colour correction.

APPENDIX A
GLOSSARY

Abbe sine condition See sine condition.

Abbe V-value A constant calculated by the ratio of a materials refraction at 587.6 nm, to its dispersion (determined by the difference in refractive index at 486.1 and 656.3 nm). See Appendix C.

aberration Ignoring the effects of diffraction (q.v.), a lens that forms an imperfect image is said to have aberrations. Aberrations can arise from lens shape, thickness, material, field position and even mechanical variations. See astigmatism, chromatic aberration, coma, distortion, field curvature, spherical aberration.

absorption Some of the energy carried by light in an optical system can be absorbed by that system. This is usually expressed as a bulk absorptance for the material considered; in optical fibres it is usually measured as dB/km.

accommodation Accommodation relates to the human eye and refers to its ability to focus. Most comfortable viewing is achieved when the eye is relaxed, that is, accommodated for an infinite conjugate. The closest focal distance is about 250 mm for healthy young adults, but it varies widely and deteriorates with age.

achromatic A lens designed to bring two different wavelengths to the same focus is said to be achromatic or an achromat. If a lens is 'corrected' at two wavelengths then it must be manufactured from a combination of two different glasses, usually a crown and a flint.

achromatization The act of correcting a lens to operate at two independent wavelengths.

acuity *Visual* acuity is the ability of the human eye to perceive fine detail in an object. For a young and healthy individual it is generally about 7 cycles/mm at a viewing distance of 250 mm (equivalent to 1′ of arc). *Vernier* acuity is the ability of an observer to align two lines. Some individuals are capable of achieving alignment to 1″ or 2″ of arc.

adaptation The ability of the eye to adapt to different lighting conditions by the expansion or contraction of the iris, together with photochemical changes within the photoreceptors themselves. Full adaptation covers a dynamic range of approximately 10^6. Pupil change accounts for a 10:1 variation, the remainder coming from bleaching of the photoreceptors.

afocal Of a system, one in which both object and image are at infinity. Common afocal systems include beam expanders and telescopes.

Airy disc An effect produced by diffraction at the aperture of a lens when it forms an image of a point source. It comprises a bright central disc surrounded by alternate light and dark rings. The intensity of each light ring falls with increasing radial distance from the centre.

ambient light The uncontrolled background level in a radiometric or photometric measurement, effectively noise.

Amici prism The Amici or roof prism deflects an input beam through 90° and produces an image inversion.

anamorphic Of a lens or system, one that has a different magnification in two orthogonal axes. A cylindrical lens is a good example of an anamorph. Anamorphic systems or lenses can be used in the production of astigmatic lines, as employed in some autofocus devices or for wide-screen cinema projection.

Anastigmat Of a lens, one that exhibits low values of astigmatism together with a relatively flat field. Examples of anastigmats are the Cooke triplet, Tessar and Biotypic lenses.

angles of incidence, refraction and reflection Angles measured between the surface normal and a ray of light.

angstrom (Å) 10^{-10} metres

anomalous dispersion The form of the dispersion graph is usually well behaved over the visible region, but some glasses are subject to a form of discontinuity in the refractive index function and thereby exhibit anomalous dispersion.

antireflection coating A coating applied to an optical surface to minimize reflection from that surface. An antireflection coating may comprise a single layer or many layers in a stack. Notch, or V, narrow-bandwidth coatings can be manufactured for monochromatic systems (lasers).

aperture The diameter of the largest beam of light that can enter an optical system.

aplanatic An aplanatic lens is one that is corrected for spherical aberration (q.v.) and obeys the sine condition (q.v.).

apochromatic An apochromatic lens is colour-corrected at three wavelengths and requires at least three different glass types. Common apochromats include enlarging lenses and some high-quality microscope objectives.

apodization One of two different image enhancement techniques. The first is that of enhanced high-frequency response achieved through a central obscuration of the lens aperture. A reduced transmission at the edge of the aperture will improve low-frequency response. The second technique is that of applying a variable transmission screen at the lens aperture to create a uniform image intensity. The screen effectively compensates for low transmission at the field edge.

apostilb (asb) A unit of luminance given by $1/\pi\,\mathrm{cd\,m^{-2}}$.

aqueous humour The liquid that fills the gap between the cornea and the lens of the human eye.

aspheric surface A surface that is not spheric, as described by a power series expansion that determines the departure from the base radius. Such surfaces can be useful in the control of spherical aberration or in the creation of high-aperture systems such as single-element condensors.

astigmatic lens A lens that focuses a distant point as two distinct lines that are orthogonal and separated by a small distance in the image field. Best focus of such a lens is defined as a circle of least confusion between the two line foci. These lenses can be used to correct the natural astigmatism of some laser diodes and as a basis for some autofocus systems.

astigmatism An aberration that causes a separation of the sagittal and tangential images (qq.v.).

athermalization A combination of mechanical compensation and material choice used to negate the change in the focal length of a lens due to a change in temperature. Such a change has two effects on a lens: the refractive indices of the elements change and the mechanical positions alter because of the thermal expansion of the mounting materials. The combined effect is a change in focal length. Athermalization is most commonly employed in infrared lenses for use beyond 2 to 3 μm.

attenuation The optical power loss per unit length through an optical material. It is the sum of absorption and scattering losses. Some manufacturers' data includes Fresnel reflection losses at the surfaces while others exclude this loss. The attenuation due to transmission alone is usually quoted as internal transmittance (T_i). The attenuation coefficient is wavelength-dependent.

autocollimation This is achieved when a point source is projected to infinity, then reflected back through the same lens to form a point image at a measuring graticule. The technique can be applied to measure the straightness of an optical bench.

axis, optical The line about which the symmetrical elements of a lens are centred.

back focal length (BFL) The distance from the apex of the last surface of a lens to the focus when tested in collimated light. Sometimes called the back working distance.

bandpass filter A filter that transmits light within a defined wavelength range and blocks light outside it.

barrel distortion See distortion.

beam diameter or waist The diameter is the distance between the $1/e^2$ peak power positions. For a laser this will be measured at the exit aperture of the cavity or at a beam focus.

beam expander A lens, usually afocal, for increasing the diameter of a collimated beam.

beamsplitter A device for dividing an optical beam into two or more optical paths. Beamsplitters can be of the cube type, manufactured from two right-angle prisms cemented together, the solid glass plate type or the pellicle type, which uses a thin plastic membrane. Splitting proportions can vary from 95 per cent through to 5 per cent, assuming a perfect transmission. Beamsplitters can also be used to produce two orthogonally polarized beams. A microscope slide can make a useful beamsplitter. See cube beamsplitter, pellicle beamsplitter, plate-type beamsplitter.

biconcave, biconvex A biconcave lens is thinner in the middle than at the edges. A biconvex lens is thicker in the middle than at the edges.

birefringence A property of some materials that causes an incident randomly polarized beam of light to separate into its two orthogonal polarization states, the ordinary (o) and extraordinary (e) rays.

blackbody radiation Radiation emitted from a thermal source with unity emissivity. The emitted radiation is characterized by the temperature of the source, often referred to as colour temperature. As a blackbody emits light in all directions it is also a Lambertian source.

blur circle See depth of focus.

borosilicate A glass type with a high resistance to thermal shock. It is often the material for the first lens in a condensing system because of the proximity to the hot source.

Bragg cell A series of acoustic transducers which produce a bulk acoustic standing wave pattern in certain crystals, e.g. lithium niobate. This pattern results in a periodic refractive index variation that can diffract an incident beam of light.

Usually the cell is arranged to maximize the power in the first diffracted order, which is also frequency-modulated at the same rate as the acoustic transducers. The Bragg cell can be used as an optical switch or as an FM modulator in heterodyne interferometers.

Brewster angle If light is incident at a planar refractive index boundary at the Brewster angle then the reflected beam is linearly polarized and contains about 15 per cent of the incident power. The transmitted beam remains randomly polarized. The polarization vector of the linear beam lies within the plane defined by the surface normal and the beam direction.

brightness The brightness of a source is a constant defined by its power per unit area and its unit solid angle of emission. A variety of units are employed but the standard SI unit is $cd\,m^{-2}$.

calcite crystal A birefringent crystal used in the manufacture of polarizing prisms.

candela (cd) The SI unit of luminous intensity, given by $lm\,sr^{-1}$. A point source with a luminous intensity of 1 cd radiates 1 lm into a solid angle of 1 sr. It is exactly defined as 1/60th the normal intensity of 1 cm^2 of a blackbody at the solidification temperature of platinum.

cardinal points The cardinal points of a lens include the positions where the focal, nodal and principal planes intersect the optical axis. Object and image distances measured to these points allow the thin-lens equations to be used.

cassegrain An optical system that consists of two mirrors that face each other. The first, or primary, mirror reflects light back towards the secondary mirror. The latter then forms an image beyond the primary, which has a hole in it to allow the light out. Cassegrains are used to shorten the optical path in astronomical telescopes and microscope objectives. As with all mirrors the design is free from chromatic aberration.

catadioptric A catadioptric lens comprises both reflecting and refracting elements. This combination can be used to good effect when the reflecting or mirror surfaces contain most of the power of the system and the refracting element(s) little or no power, so producing images of good quality.

catoptric The catoptric lens is a mirror-only system, where all the optical power is provided by reflecting surfaces.

charge-coupled device (CCD) A regular array of detectors that integrate charge arising from the incidence of light. They come in two guises: first as a linear array of elements and second as an area array. The CCD is now a common imaging sensor and is used in many closed-circuit television (CCTV) applications.

chief ray The ray originating at the extreme object height and passing through the centre of the entrance pupil; sometimes called the principal ray.

chroma Chroma or saturation is concerned with the purity of a colour, effectively the amount of white that it contains.

chromatic aberration Errors in the performance of a lens which arise from differences in focal length and lens power caused by variation in refractive index with colour.

chromaticity coordinates Chromaticity coordinates are a means by which a colour can be represented mathematically on a chromaticity diagram. The most common chromaticity diagrams are the 1931 CIE (x–y) and the 1976 (u'–v') uniform chromaticity diagrams.

chromaticity diagrams See chromaticity coordinates.

CIE The Commission Internationale de l'Éclaitage (CIE) is the international commission on illumination. It is the body responsible for international agreement on the mathematical treatment of colour data.

circle of least confusion The region of best focus of an aberrated system. This term applies in particular to astigmatic systems where the circle of least confusion lies between the sagittal and tangential focal lines.

circularly polarized light Light in which the polarization vectors are out of phase by a quarter of a wavelength.

cladding The lower refractive index material that surrounds the core of an optical fibre.

coherence length The distance over which coherence is maintained. It is the limiting path difference between the arms of an interferometer.

coherent bundle If the fibres of an optical bundle are carefully arranged so that the image at one end of the bundle is faithfully reproduced at the other end then it is said to be coherent.

coherent radiation Coherent radiation is that in which a constant phase relationship is maintained between all points on the wavefront. This applies both spatially and temporally.

cold mirror A cold mirror reflects visible light and transmits the infrared radiation which causes materials to heat up.

collimation Collimated light is established when all the rays of a beam are parallel to each other.

colour temperature A quantity used to characterize the spectral output of a light source. If the chromaticity coordinates of a light source match those of a perfect blackbody, then the source can be said to have colour temperature equivalent to the absolute temperature of that blackbody.

coma This aberration arises from the difference in magnification as a function of aperture and causes a comet-like tail on the image of a point source. Normally coma is confined to off-axis images, unless centring or tilt errors are present.

condenser Condensing optics are used to collect light from a source. Condensers can be mirrors or lenses. A typical lens condenser might be aspheric in form to maximize light-gathering ability (low f-number) and manufactured from a borosilicate glass to withstand potentially high source temperatures. A mirror-based condenser can take the form of a deep elliptical reflector that collects light over a large solid angle.

cones The colour-sensitive photoreceptors in the human eye.

conjugates The distances from the image and object positions to the principal points of the lens.

contrast transfer function (CTF) The square-wave spatial frequency amplitude response of an optical system. CTFs cannot be combined together to give the overall response of a system.

core The high refractive index region at the centre of an optical fibre, surrounded by the cladding, which conducts the light.

cornea The transparent, hard outer casing of the human eye that supplies most of its refracting power.

correlated colour temperature If the radiation from a light source does not follow the Planck equation it cannot be assigned a colour temperature. In this case, a correlated colour temperature is used, to provide the closest match between the actual source colour and a blackbody. Correlated colour temperature can be determined graphically or mathematically.

cosine laws The variation of luminance as a function of source type, together with the angle of light from the source and the angle received by the detector. See Appendix C.

critical angle The lowest angle of incidence at which total reflection occurs at the boundary between two media of different refractive indices.

cube beamsplitter A device formed by two right-angle prisms cemented together along their hypotenuse faces. One of the hypotenuse faces has deposited on it a partially reflecting, partially transmitting coating that operates efficiently at incidence angles of 45 °.

cylindrical lens Anamorphic lenses with surface curvature in one plane only.

dark adaptation The ability of the eye to see in low light levels. It is based on the photochemical change in the rods, which enhances their sensitivity to low light levels.

dark noise The current that flows in a photodetector in the absence of light.

decibel (dB) The decibel is a standard logarithmic measure of relative power; $\mathrm{dB} = 10 \log(P_2/P_1)$.

depth of field If a lens is focused at a particular point then the depth of field is the distance that the object can move along the optical axis before an unacceptable change in the image quality is observed.

depth of focus The region either side of the true focus where imaging performance is acceptable for the lens application.

detectivity The reciprocal of noise equivalent power (NEP) (q.v.) in units of reciprocal watts.

dichroic filters These reflect or transmit light depending on wavelength. They are often used for the control of heat within an optical system, where they are referred to as hot or cold mirrors. Other common applications include colour selection filters for TV cameras or LCD projectors.

diffraction Wavefronts do not travel in straight lines, but bend around objects and edges. This phenomenon is called diffraction and is a consequence of the wave nature of light. Diffraction is the ultimate limit to the performance of a lens.

diffraction grating A diffraction grating is a reflecting or refracting structure of lines or grooves that produce diffraction. It is best used in collimated light. When white light strikes the grating, each line element causes diffraction-creating spectra on each side of the undiffracted light, the zero order. Counting from the centre, the spectra are ascribed order numbers, 1, 2, etc. A blazed grating is one designed so that the majority of diffracted light falls in one order only. Diffraction gratings are dispersive elements with shorter wavelengths deviated through the largest angles.

diffraction-limited performance A diffraction-limited lens is one where performance, or resolution, is limited by diffraction only, not by aberration.

diffuser A material or surface that interacts with light to modify its spatial or angular distribution in a predetermined manner. Examples include ground-glass or opal screens.

dioptre The power of a lens is measured in dioptres. The dioptre rating of a lens is the reciprocal of the focal length measured in metres. Thus a lens with a focal length of 500 mm has a power of 2 dioptres. The dioptre power of a surface depends on both its curvature and the magnitude of refractive index change across it.

dispersion The splitting of light into its constituent wavelengths. Diffraction gratings and prisms are examples of dispersive elements. Glass dispersion refers to the variation of refractive index with wavelength; typically the refractive index of glass will fall with increasing wavelength. Mode and material dispersion in optical fibres causes a broadening of the input pulses along the length of the fibre limiting the bandwidth by the degree of pulse spreading.

distortion An aberration that causes the image size and therefore magnification to vary with field position. Positive, or pincushion, distortion transforms a square object into an image with elongated corners. Negative, or barrel, distortion produces an image with bulging sides.

divergence Even the best-quality laser beam is not truly collimated; as it moves away from the cavity it will spread out. This inherent divergence of a laser beam is due to diffraction and is typically measured in milliradians.

doppler effect The apparent change in frequency due to the relative motion of source and detector; thus light moving away from an observer is downshifted in frequency.

double Gauss Usually a photographic objective lens, which in its simplest form consists of two negative meniscus doublets inside two positive lenses, all symmetrically placed about the stop.

double refraction See birefringence.

doublet A compound lens made from two elements, usually of different glass types, which work together to compensate for aberrations, usually chromatic. The lens can be air-spaced or cemented.

Dove prism A truncated right angle prism that exhibits the characteristic of rotating its image at twice the rate at which the prism can be rotated. The prism returns best performance when operated with collimated light.

dye laser A laser type which uses a dye as its lasing medium; typically these lasers have a broad spectrum and are often used in ultra-short-pulse-length applications.

edge filter A filter that transmits light above or below the wavelength defined by the edge position, e.g. a blue or red filter.

edge-light-emitting diodes (ELEDS) Most LEDs are surface-emitting, however a few are produced in which the light emission occurs from the side of the device. These are particularly appropriate for the communications industry, where bonding to fibre optics or fibre ferrules is important.

effective focal length (EFL) In a lens this is defined as the distance from the second principal plane to the position where a paraxial bundle of collimated rays come to a focus. For a lens with negative power the divergent ray bundle is extrapolated back to focus, and again the distance between this virtual focus and the principal plane is the EFL of the lens.

electro-optic effect The change in refractive index caused by the application of an electric field.

elements A collection of lenses that performs a particular function within a complex lens is called a group. Each lens of a group is called an element.

emissivity The ratio between the radiance from a light source and that from a blackbody at the same temperature and wavelength.

endoscope Usually a medical instrument, this is often used to view inside the human body. Modern endoscopes usually contain coherent fibre optic bundles but can also contain lens systems.

entrance pupil The image of the stop as seen from the object side of the lens.

equibiconvex An equibiconvex lens is a singlet that has equal and opposite curvature on each of the two faces such that a positive lens is formed (real focus). The lens is thick in the middle and thin at the edges.

etalon A system, usually a pair of mirrors, that can bounce light back and forth. Both mirrors may be plano, or one or both may be curved so as to form a focus. With the correct choice of separation between two plane mirrors a wavelength-selecting device can be made. In this case the cavity size is a whole number of wavelengths.

étendue The flux per unit radiance or luminance of a light beam.

excimer laser Ultraviolet-emitting lasers characterized by short, energetic pulses.

exit pupil The image of the stop of a lens as seen from the image side.

exitance The exitance of a source is defined as the flux leaving its surface per unit area.

extraordinary ray In a birefringent material the extraordinary ray does not necessarily obey Snell's law on refraction at the crystal surface.

eye relief This is the distance from the vertex of the last surface of an eyepiece to the exit pupil of the optical system.

eyepiece An eyepiece, or ocular, is the final lens of a visual system which magnifies a real intermediate image and presents it to the observers eye. Common types are the Huygens, Kellner, Ramsden and Erfle.

f-number (*f*#) A measure of the light-gathering ability of a lens; it is only truly applicable when considering one infinite conjugate. The f-number of a lens can be determined by dividing the focal length of a lens by its clear aperture (diameter).

Fabry–Pérot interferometer A device constructed from two plane parallel plates with high reflectivity. Light for which the cavity length is a whole number of wavelengths and for which the propagation direction is normal to the plates constructively interferes. An example of this is a laser cavity resonator.

fibre optic This term is applied to a wide variety of fine glass or plastic light guides. Fundamentally, a fibre optic is a thin strand of glass that will support light propagation. The fibres can be singular, as for data communication, or bundled together.

field angle The angle subtended by the object at the centre of the stop relative to the optical axis.

field curvature An abberation at the image plane of a lens which makes the plane in fact a curved surface. A lens corrected for this aberration is said to have a flat field.

field lens The action of a field lens is to increase the field of view without increasing power or magnification. It is usually placed at or near an intermediate image such that extreme rays are bent into the exit pupil. Field lenses are most often used within telescopes, in sighting systems and in radiometric applications.

fluorescence The emission of longer-wave radiation following stimulation by a shorter wavelength. It persists for less than 10^{-8} s.

fluorescent sources The fluorescent light source is a low-pressure discharge lamp with phosphors coated on the inside of the envelope that are activated by the uv light from the discharge. Depending on the blend of the phosphor and type of gas fill, a variety of spectral outputs are available.

focal length See effective focal length.

footlambert (fL) The footlambert is a unit of luminance and is equal to $1/\pi\,\mathrm{cd\,ft^{-2}}$.

Fraunhofer diffraction Fraunhofer diffraction is concerned with the far field, that is, diffraction observed far from the diffracting aperture or edge. Fraunhofer diffraction caused by a lens aperture appears in the image plane and degrades the image. It is this diffraction that limits the performance of an optical system.

Fresnel diffraction An effect that is observed close to the diffracting aperture or edge, and is often produced with a point source at a finite distance. It is also called near-field diffraction.

Fresnel lens A Fresnel lens is typically made up from a number of fine annular facets. The facets can be closely spaced, as for an overhead projector condenser, or coarsely spaced as for a theatre spot projection lens. Fresnel lenses are also used as field lenses, for TV projection and with security sensors, or for any application where low cost and low weight are important.

Fresnel reflection When light is incident on an air–glass interface most of it is transmitted but some, about 4 per cent typically, is reflected. This reflection occurs when light is incident at any refractive index boundary. The amount of reflected light is governed by the difference in refractive indices and the angle of incidence. See Appendix C.

front focal length The front focal length of a lens is the distance between the front focus position and the first principal point.

full-width half maximum (FWHM) The full width at half maximum is a convenient and near-standard way of characterizing a well-behaved photometric distribution such as a Gaussian or cosine. The full width of the distribution is measured at half the height of the peak. This definition is also applied to bandpass filters.

Galilean telescope A telescope, attributed to Galileo, comprising of one positive element with its focus coincident with the virtual focus of a negative element. A collimated beam is either expanded or compressed through this telescope, which is afocal (no focus, collimated in and collimated out). The telescope has no internal foci and is therefore ideal for medium-power laser beam expansion, where beam focusing could cause local ionization problems.

Gaussian beam A light beam whose intensity distribution is described by Gaussian statistics, so that beam width is defined by the $1/e^2$ points.

Gaussian optics See paraxial optics.

geometric optics Ray tracing that ignores the effects of diffraction and assumes that ray bundles travel in straight lines.

ghost image The image formed by unwanted reflections within a lens system. Usually most obvious when uncoated optics are used.

Glan-Thompson prism See Wollaston prism.

glare See stray light.

gloss That property of a surface that maintains a specular reflection at low or grazing angles of incidence. A surface of high gloss is not necessarily a mirror surface.

graded-index fibre A type of optical fibre in which the refractive index of the core decreases radially from the axis toward the fibre cladding.

GRIN lens A (*gr*adient *in*dex) lens in which the refractive index decreases across its radius or along its axis. Consequently, it has the properties of a normal spherical lens except that the output faces are plane and parallel to each other.

ground-glass screen A sheet of glass in which one face has been ground or etched to produce a diffusing surface.

group A collection of lenses that perform a particular function within a complex lens.

half-wave plate An optical element, manufactured from a birefringent material, so that it retards the phase of the slow ray with respect to the fast ray by half a wavelength. It is commonly used for rotating the polarization state of a laser.

heat-absorbing filter An optically clear element used to absorb infrared radiation from a beam of light thus limiting subsequent optical elements from heating up.

helium-neon (He-Ne) laser A widely used gas laser with many different applications. Its common operating wavelength is 632.8 nm (red), but variants are commercially available in the green, orange and infrared.

heterodyne Usually applied to detection systems that combine two frequencies to generate a third that is equal to the difference between them.

hologram An interference pattern recorded on a high-resolution emulsion or polymer. This complex phase grating can be used in place of many optical elements and finds its most common application in head-up displays (HUDs) where low weight is an advantage.

holography The interference of two coherent beams, one is a reference and the other illuminates the subject. When these two beams are combined, the resultant interferogram contains phase information about the subject. The interferogram can be recorded as a hologram, which can be used to recreate an image of the original subject by illumination with the reference beam.

hot mirror A mirror that transmits the visible light and reflects the heat, producing infrared radiation.

hue The property that allows colour distinction, for example blueness, redness or greenness. Achromatic colours, such as black, white or grey, do not exhibit hue.

Huygens principle This states that each point on a wavefront acts as a secondary source of wavelets. The amplitude of these wavelets can be summed to generate the next position of the wavefront.

hyperfocal distance The object distance at which a projection lens must be focused to give the maximum depth of field, extending from half the hyperfocal distance to infinity.

hyperopia The reverse of myopia; sometimes called longsightedness or farsightedness. It occurs because either the eye is too short or the lens power is too weak, resulting in distant images being focused behind the retina.

illuminance The visible optical flux per unit area falling on a surface. SI unit: lx, or lm m^{-2}. Sometimes called illumination or luminous flux density.

incandescent lamp A lamp that emits light when the filament is heated by an electric current, a common example being the tungsten–halogen lamp.

incoherent radiation Light emitted from a discrete source with random phase relationships.

infrared The infrared spectrum encompasses every wavelength longer than those in the visible (beyond about 770 nm) and shorter than microwaves (about 1 mm). For imaging through the atmosphere the infrared is sometimes classified in three regions. The near-infrared encompasses those wavelengths from 770 nm to about 1.5 μm, the middle infrared from about 3 to 5 μm and the far infrared from about 8 to 12 μm.

insertion loss A loss incurred when an optical fibre is connected to a system. It is typically due to reflection or misalignment and measured in decibels.

integrating sphere A spherical matt reflector used to measure the total flux from a light source.
intensity Flux per unit solid angle. Measured in cd (lm sr^{-1}) or W sr^{-1}.
interference Light, or any electromagnetic radiation, of the same wavelength that combines or adds is said to interfere. Light in phase that adds gives interference that is constructive. Light out of phase that cancels gives interference that is destructive.
interference filter An optical filter manufactured from a number of vacuum-deposited dielectric layers, which together form an etalon resulting in a narrow-band filter typically of 5 to 50 nm full width at half maximum (FWHM).
interferometer An optical instrument that combines two coherent optical beams so that an interference pattern is generated. The instrument is generally used for testing the quality of optical components.
internal transmittance The light transmitted through a material excluding Fresnel reflection losses at the surfaces.
inverse-square law This states that the measured intensity will fall in inverse proportion to the square of the distance from the emitting surface.
irradiance Irradiance is the radiant flux density falling on a surface. The SI unit is the W m^{-2}. The photometric quantity is illuminance or luminous flux density at a surface, of which the SI unit is the lm m^{-2} or lx.
Ishihara test A test for colour blindness or colour deficiency.
Kerr cell Some materials exhibit birefringence when placed in a strong electric field. This effect can be used to create a Kerr cell by placing the material between crossed polarization analysers, so creating a very fast optical shutter.
Lagrange invariant A quantity relating to the passage of a single paraxial ray through an optical system. It can be defined as the product of object height, paraxial ray angle and the refractive index of the object space. This product remains unchanged whether in object or image space. Also called the optical constant or Helmholtz invariant.
lambert (L) A unit of luminance equivalent to $1/\pi$ cd cm^{-2}.
Lambertian emitter An optical source whose radiance is constant in all directions.
laser A *l*ight *a*mplification by *s*timulated *e*mission of *r*adiation. The laser is an intense source of monochromatic and coherent radiation.
laser diode The conventional light-emitting diode emits light spontaneously at a semiconductor junction. In the laser diode the current is raised above a certain threshold to cause optical feedback to take over and stimulated emission to occur. In the semiconductor laser, just like in the gas laser, population inversion is required, plus some form of optical cavity within which to contain and control the laser emission.
lateral colour An aberration that produces a variation in image height as a function of wavelength.
LED A semiconductor device that emits light when forward-biased.
longitudinal colour An aberration that produces a variation in focal length as a function of wavelength.
lumen (lm) The SI unit of luminous flux or photometric power, analogous to the radiometric watt. It is the luminous flux emitted per steradian by a standard point source with a luminous intensity of one candela, i.e. 1 lm = 1 cd sr^{-1}.
luminaire Any form of light fitting designed and manufactured specifically for the

supply of light. Luminaires may take the form of emergency lights, decorative fittings for the home, street lighting or accent lighting.

luminance The luminous flux per unit solid angle measured normal to the direction of propagation. Its SI unit is the nit (nt), or cd m^{-2}.

luminous efficacy Unfortunately the quantity known as luminous efficacy can be defined in two ways, both of which are in common usage:

1. The ratio of luminous flux emitted by a light source to its input electrical power. It has the SI unit of lm W^{-1} (electrical).
2. The ratio of total luminous flux to the equivalent radiant flux emitted in the photopic region. The unit is the lm W^{-1} (radiant); the quantity is used to convert flux in watts to flux in lumens. The maximum value of luminous efficacy, for the photopic region, is 683 lm W^{-1} and occurs at 555 nm. For scotopic applications the maximum value of luminous efficacy becomes 1700 lm W^{-1} at 510 nm.

luminous efficiency Usually expressed as a percentage and, using the second definition of luminous efficacy (q.v.), it is the luminous efficacy divided by 683 (photopic) or 1700 (scotopic).

luminous flux density See illuminance.

luminous intensity This is the luminous flux within a given solid angle. The unit of luminous intensity is the cd (lm sr^{-1}). For a given luminous flux (q.v.), measured in lm, the luminous brightness is controlled by the solid angle into which the source emits.

lux (lx) The SI unit of illuminance or luminous flux density, equal to 1 lm m^{-2}.

magnification, angular The ratio of the angles subtended respectively by the object and image at the principal planes of an optical system.

magnification, lateral For systems entirely in air, this is most commonly defined as the ratio of image height to object height. It is also defined as the ratio of image distance to object distance, measured from the principal planes of a system.

magnification, longitudinal The magnification of the longitudinal thickness of the object, which for small thicknesses is equal to the square of the lateral magnification.

Mangin mirror This mirror comprises a lens with a mirror surface deposited on one side. This provides a reflecting element with the potential for correction of spherical aberration and finds common application as the secondary mirror of reflecting telescopes.

marginal ray The marginal ray of an optical system is the paraxial ray that passes through the extreme of the aperture stop.

meniscus lens A singlet lens having both surfaces with curvature of the same sign. The surfaces can be both positive or both negative.

meridional plane See tangential plane.

mesopic vision The intermediate fashion in which the eye behaves between the extremes of scotopic vision and photopic vision.

metamerism Quantities of light with different spectra but having the same colour coordinates are said to exhibit metamerism.

micrometre (μm) An SI unit of length equivalent to 10^{-6} m.

micron (μm) Old name for micrometre (q.v.).

mie scattering The scattering from particles that are approximately the same size as the wavelength of the incident radiation.

modal dispersion Pulse spreading caused by differential path length in a multimode optical fibre. Different rays travel different paths within the multimode fibre, and hence some rays have an optically longer path to follow. This results in a distribution with respect to the time of arrival at a detector.

mode The standing wave pattern supported by a laser cavity or propagated by an optical fibre.

modulation transfer function (MTF) The sinewave amplitude response of a lens or other optical system. It is a fundamental measure of a lens's resolving capability and is typically displayed as a graph of contrast versus spatial frequency in the image plane.

Moiré patterns Moiré fringes are produced by either the interference or the shadowing of two grating patterns. With a suitable detection system Moiré fringes can be used for accurate positioning of machine tools and in encoders.

monochromatic light Light containing radiation of a single wavelength only. In practice monochromatic light sources will have a finite bandwidth.

monochromator An optical instrument that separates light into narrow bandwidths usually achieved using a dispersive element such as a prism or grating.

monomode fibre optic Monomode fibres support only one propagation mode. This is achieved by reducing the diameter of the core until only one mode is left. Typically this diameter will be of the order 5 μm. The fibre is of the step index sort. The cladding of a monomode fibre has to be quite thick, because much of the energy propagated by the fibre will travel as an evanescent wave within the cladding. Monomode fibres exhibit no modal dispersion.

multilayer coating This is an optical coating that can comprise many layers of materials, with alternating high and low refractive index. The simplest form is the two-layer design, with the first layer of low index and the second of higher index material. Each layer has an optical thickness of a quarter-wavelength, so both layers introduce a 180° phase shift at the air interface. Destructive interference occurs and the reflection is minimized. Applications include antireflection coatings and wavelength selection such as coloured filters, heat rejection or reflection.

multimode fibre optic The multimode, or graded index, fibre has a radially symmetric refractive index profile with the highest index being along the axis of the fibre.

myopia Commonly referred to as short sightedness, this eye defect forms an image of a distant object in front of the retina. This defect is caused by the eye being too long or the eye lens too powerful.

nanometer (nm) An SI unit of length equal to 10^{-9} m.

neutral density filter An optical filter with a uniform attenuation across a defined waveband. These filters are usually graded according to optical density, which is the logarithm to the base 10 of the inverse of internal transmittance.

Newtonian telescope A reflecting telescope with a parabolic primary and a plane steering mirror, together forming an image outside the telescope tube for viewing by an eyepiece.

Newton's rings A series of bright and dark rings formed by the interference of light between two optical surfaces separated by a small air gap. Commonly used in optical work shops where a test plate is placed in contact with the surface to be inspected.

Nicol prism See Wollaston prism.

nit (nt) The SI unit of luminance (brightness); equal to 1 cd m^{-2}.

nodal points The nodal points of an optical system lie on the optical axis of the lens. A ray directed at the first nodal point appears to emerge from the second nodal point without a change in direction.

noise equivalent power (NEP) The ratio of noise current to photodetector sensitivity at a specified wavelength and frequency range. Measured in W.

numerical aperture (NA) A measure of the light-gathering ability of an optical system. It is the finite conjugate f-number and is therefore linked to resolving capability. Numerical aperture is also a measure of the light accepted by a fibre optic.

objective The first element of an optical system that receives light from an object and forms the primary image.

optical blacking A finish applied to optical surfaces to absorb light and reduce stray light reflections.

optical cement Any adhesive used in joining optical elements. Optical cements are selected for their ability to minimize Fresnel reflection losses between surfaces; common examples include Canada balsam and UV-curing epoxy.

optical path The optical path of a ray is defined as the product of distance travelled and the refractive index of the medium. The optical path is integrated over regions of varying refractive index.

optical path difference (OPD) The departure of a real wavefront from the ideal, or reference spherical, wavefront is often measured in terms of optical path. The magnitude of departure is the optical path difference measured in wavelengths.

optical waveguide Any material structure capable of guiding radiation along a path parallel to its axis containing the light within its boundaries or adjacent to its surface.

ordinary ray In a birefringent crystal the ordinary ray is the one that obeys Snell's law on refraction at the crystal surface.

paraxial optics Paraxial, or Gaussian, or first-order, optics are a simplification of geometric optics in which rays travel close to the axis and apertures are considered as infinitesimally small. This allows analysis using thin-lens approximations and ignores the effect of aberrations.

parfocal Optical systems that are parfocal can be interchanged without the need to refocus. Common parfocal lenses are microscope objectives, eyepieces and CCTV or CCD lenses.

pellicle beamsplitter The pellicle beamsplitter is a thin, elastic membrane (the pellicle) stretched over a frame. This type of beamsplitter offers the major advantage of virtually eliminating the second surface reflection by making it nearly coincident with the first. The pellicle is very light, which is useful in certain applications.

penta prism Penta prisms turn an image through 90° without altering the image orientation. The penta prism has two reflective coated surfaces because some landing angles will not exceed the critical angle for total internal reflection (q.v.).

Petzval surface A curved surface on which the image lies without exhibiting astigmatism.

phosphorescence The emission of light delayed by more than 10^{-8} s following stimulation.

photodiode A reverse-biased p-n junction. Photons incident on the photodiode liberate electron–hole pairs in the region of the p-n junction. This results in the lowering of the p-n potential barrier, so that current flow can occur. The more numerous the photons are, the greater is the current that flows.

photoemission This occurs when photons incident on an absorbing surface are sufficiently energetic to enable the liberation of electrons from that surface. This effect is used as the basis of detection in certain UV sensors.

photometry The measurement of radiation in the visible region, weighted by the photopic response.

photopic vision The normal light-adapted operation of the human eye, specifically within the visible waveband to which the cones in the human eye respond.

pixel A condensed form of 'picture element'. Primarily, the small, photosensitive cells of a linear or area CCD used for image recording.

Planck's law This associates the quantum nature of photon energy to its frequency or wavelength. It states that the photon energy is directly proportional to its frequency. The constant of proportionality is the Planck constant.

Planck's radiation law This defines the spectral radiant exitance of a blackbody in terms of its absolute temperature and the wavelength of the emitted radiation.

plane-polarization Light is plane-polarized when the electromagnetic vectors all vibrate in the same plane.

planoconcave A singlet lens with one flat surface and one curved surface. The lens is thin at its axis and thick at its edge (caves go inwards).

planoconvex A singlet lens with one flat surface and one curved surface. The lens is thick at its axis and thin at its edge.

plate-type beamsplitter A solid plate of optical material, usually coated, so as to provide a beamsplitting function at a particular angle of incidence.

Pockels cell An intensity modulator that comprises a birefringent crystal between a pair of crossed plane polarizers. The application of a varying electric field in the crystal causes a rotation of the output polarization vector and results in an intensity modulation of the light leaving the second plane polarizer.

Pockels effect An effect found in some birefringent materials where the degree of relative phase retardation between the 'o' and 'e' rays is directly proportional to the applied electric field strength.

point source The definition of this depends on the context of its use. In lens design a point source is infinitely small and all rays originate from the same position. In lens testing a point source is one that is smaller than the resolving capability of the lens under test. In radiometry or photometry a light may be considered a point source if the distance from it to the detector is greater than about ten times its longest dimension.

point spread function The point spread function is the intensity distribution across a diameter of the image of a point source. It is a measure of the quality of the optical system producing the image.

polarization Light propagation is the result of the alternate interaction between electric and magnetic fields that are perpendicular to each other. The polarization state of light is defined by the direction of the electric vector. The polarization vector can be resolved into two orthogonal components, which usually travel at the same speed, perpendicular to the direction of motion. Ordinarily, light is unpolarized with equal amounts of energy in the two polarizations (randomly polarized). If only one of the resolved polarization vectors is present then the light is plane, or linearly, polarized. If a phase difference exists between the two resolved components the light is either elliptically polarized, or, if the phase differs by $\pi/2$, circularly polarized. The two orthogonal polarization states are commonly referred to as the s and p states. For the p-polarization the electric field vector is parallel to the plane of incidence, the s-polarization is perpendicular to it. In a birefringent material these states may be referred to as the ordinary (o) and extraordinary (e) rays.

polarizing prism See Wollaston prism.

polychromatic Light of many different wavelengths. The converse of monochromatic, which is light of one wavelength or frequency only.

Porro prism A prism that can be considered as two smaller right-angled prisms attached so as to form one large right-angle one. It is used such that light enters at one side of the long hypotenuse face normal to the surface (thus minimizing dispersion effects). This light is totally internally reflected by the two short faces and emerges from the opposite end of the hypotenuse face. The beam is steered through 180°.

power See dioptre.

presbyopia Presbyopia is the inability of the eye to accommodate or focus. It is due to the hardening of the crystalline lens.

principal planes and points The positions in space at which the apparent bending of a light ray occurs. See cardinal points.

pulse spreading The increase in pulse width within a given length of optical fibre, or other waveguide, due to the effects of modal and material diepersion.

pyroelectric detector A common detector of 'thermal' infrared radiation. The detector is a current source whose output is proportional to the rate of change of temperature. It can be operated at room temperature (300 K) and is usable over the wavelength range 0.6 μm to 35 μm.

quarter-wave plate A birefringent material whose thickness is designed to introduce a retardation of a quarter of a wavelength between the 'o' and 'e' rays.

quartz A glass type with a relatively broad transmission spectrum, particularly used in UV and IR applications.

radian A unit of angle, there being 2π radians in 360°.

radiance The radiometric brightness of a light source, usually measured in $W\,m^{-2}\,sr^{-1}$

radiometer A device used to measure the intensity of a radiant source.

radiometry The science of measuring radiation across the whole of the electromagnetic spectrum. The branch of radiometry specific to the visible region of the spectrum is photometry.

Rayleigh limit This sets the maximum acceptable departure from the perfect spherical wavefront at a quarter of the wavelength for good imaging performance.

Rayleigh scattering Scattering from particles which are much smaller than the wavelength of the light. The scattered power is inversely proportional to the fourth power of the wavelength. This is why the sky is blue.

Rayleigh's criterion See resolving power.

reciprocal relative dispersion See Abbe V-value.

reflectivity A measure of the amount of light reflected from a surface. It may be quoted as a percentage or as a number between zero and one and is part of the expression

$$\text{reflectivity} + \text{transmission} + \text{absorption} = 1.0$$

refraction The bending of an oblique ray of light, incident on a boundary between two materials with different refractive indices.

refractive index The ratio between the speed of light in a particular material and its speed in vacuum.

resolving power A measure of the ability of an optical system to discriminate between two closely spaced objects.

responsivity A figure that indicates how many amps of current will flow from the detector for the watts of radiant power falling on it. Owing to the spectral dependence of this parameter it is usual to have it quoted at one wavelength or to supply a graph of the response. Typically, near the peak of response the responsivity of silicon should be of the order 0.5 A W^{-1}.

retardation plate A polarization component with fast and slow principal axes. Polarized light falling on the element is first resolved into two orthogonal polarizations and then recombined on exit to give a single new polarization state rotated with respect to the original. Examples include quarter wave, half wave and full wave retardation plates.

reticle An optical component located at an image plane; it is usually part of a relay or magnifying system. In this way any structure within the reticle, such as a grating or cross hairs, can be overlaid with the object being viewed. The reticle is not usually at the final image plane.

retina The light-sensitive part of the human eye responsible for the visual sensation.

retroreflector A reflecting element or system that returns light along the path of incidence; it is often composed of three mirrors that form the corner of a cube.

rhomboid prism The rhomboid prism uses TIR at two faces to displace the optical axis laterally. The double reflection creates a single-axis double inversion and therefore an erect and correctly handed image is transmitted.

right-angle prism A device commonly used to deviate rays through 90° using total internal reflection (TIR) at the hypotenuse face. Right-angle prisms deviate the ray bundle through 90° but introduce an image inversion.

Rochon prism A prism used specifically in ultraviolet applications to introduce polarization.

rods The non-colour-sensitive photoreceptors in the human eye responsible for night or dark vision when levels of illumination are low.

roof prism See Amici prism.

rouge A fine polishing material made from powdered ferric oxide. It is used in the production of optical surfaces.

Royal Microscopical Society (RMS) thread The mounting thread of a microscope objective is a standard Royal Microscopical Society (RMS) fitting and is defined according to British Standards Institution (BSI) specifications. Details of this standard thread are given in Table 2.1. (p. 47)

sag The height of a curve measured from the chord.

sagittal plane The plane that contains the optical axis but is orthogonal to the plane containing the object.

scattering The change in angular or spatial distribution of light caused by its interaction with a volume or surface.

Schmidt corrector (plate) Often used in astronomical telescopes, this is a thin aspheric element sited at or near the radius of a spherical mirror to correct for spherical aberration.

scotopic vision The activity of the human eye in its dark-adapted state.

secondary colour The residual aberration remaining after correction of primary aberrations. For example, in an achromatic design the red and blue foci are forced to coincide, but, secondary colour exists if the green focus is not at this position.

seidel aberrations The primary aberrations of an optical system. They comprise spherical aberration, astigmatism, coma, distortion and field curvature.

semiconductor laser See laser diode.

sine condition This paraxial condition must be met by lenses corrected for spherical aberration if they are also to be free of coma near the optical axis. The sine condition states that the ratio of the paraxial ray angle to the sine of the marginal ray angle is a constant in image and object space.

single-mode fibre An optical fibre used for high-grade communications. It is characterized by a small core diameter that supports only a single propagation mode and hence limits modal dispersion.

skew ray For a rotationally symmetric system a skew ray is one that does not lie within a plane containing the optical axis.

Snell's law The law that describes the bending of light at a refractive index boundary. It is the fundamental basis of all ray tracing. The mathematical form is presented and discussed in Chapter 1.

solid-state laser A laser comprising a solid rod or crystal, such as Nd–YAG, which is the lasing medium stimulated by a pump source such as a flashlamp or laser diode.

speckle effect, speckle pattern An interference effect arising from the spatial coherence of an expanded laser beam. It is seen as an intensity modulation, or speckle, when the beam is viewed.

spectral irradiance The irradiance per unit wavelength. Usually measured in $\mathrm{W\,nm^{-1}}$.

spectrum The wavelength distribution of radiation from a source.

specular Of reflection, mirror-like.

spherical aberration The variation of focal length with aperture; it can be defined in both longitudinal and transverse senses.

spherochromatism A chromatic variation of spherical aberration.

spherometer A device used to measure the curvature of a spherical surface.

spot diagram A graphical representation of the landing heights of rays in a transverse plane, usually near focus.

standard illuminant A light source, whose spectral distribution is not necessarily achievable in practice but has been defined to set a standard for colorimetric calculations. Examples of such sources include:

standard illuminant A A blackbody light source with a correlated colour temperature of 2856K. Practically achieved using a tungsten-halogen filament lamp.

standard illuminant B An approximation to the spectral distribution of direct sunlight with a correlated colour temperature of 4874K. This standard is no longer in use.

standard illuminant C The spectral distribution of average daylight with a correlated colour temperature of 6774K.

standard illuminant D_{65} A daylight source with a correlated colour temperature of 6504K. Most commonly used to describe a 'white light' source, especially in the TV and film world.

star test A test in which the quality of the image of a point source is judged. The test can reveal important information with respect to the quality of the optical system producing the image. In practice it is usually carried out by eye with an experienced optical worker taking the quality decision.

Stefan–Boltzmann Law The law that relates the total radiant power of a blackbody to the fourth power of its absolute temperature.

step index fibre An optical fibre whose core has a uniform refractive index up to the inner edge of the cladding.

steradian (sr) The SI unit of solid angle. It is the ratio of the area on a spherical surface intercepted by a cone to the square of the radius of the sphere. There are 4π sr in a sphere.

stereo vision Depth perception arising from the finite spacing between the eyes allowing each to see a slightly different view.

stilb (sb) The CGS unit of luminance, equivalent to 1 $cd\,cm^{-2}$.

stray light Light arriving at the focal plane other than from the design intended optical path. Stray light reduces the contrast of an image or signal-to-noise ratio at a detector and arises from reflections within the optical system.

Strehl ratio The ratio of the actual illumination at the centre of the Airy disc to that obtained from an aberration free system. This is a measure of quality of the optical system. A Strehl ratio of 80 per cent corresponds to the Rayleigh criterion of a quarter-wave of optical path difference.

stria Within an optical material, a layer, or part of a layer, or a streak that will cause a degradation to the transmitted wavefront. Usually associated with quality material specifications. Striae are regions of slightly different refractive index caused by imperfect mixing of the melt during glass manufacture.

talbot The SI unit for the quantity of light or luminous energy and equals 1 lm s.

tangential plane The tangential (meridional) plane is the one that contains the object and the optical axis.

telecentric lens One in which the aperture stop of the lens is positioned at one of the focal points, so that the exit or entrance pupil is located at infinity. This has the effect of making the chief ray parallel to the optical axis, so that the system magnification is tolerant to a change in focus. This technique is most often used in metrology.

telephoto lens A lens whose focal length is longer than its overall length. Telephotos are commonly employed as photographic lenses where a long focus is achieved with a relatively short lens to film distance. Both principal planes of a telephoto are in front of the lens.

telescope An afocal optical instrument used for viewing distant objects.

thick lens A lens in which the thickness of the glassware has an effect on the accuracy of first-order analysis. A thick lens has a finite separation between principal planes.

thin films See interference filters, multilayer coating.

thin lens A lens whose thickness can be ignored during first-order analysis without affecting the accuracy of the calculation. A thin lens will have coincident principal planes.

toric lens A lens in which the curvatures in two orthogonal planes are different.

total internal reflection (TIR) Total internal reflection occurs when light passes from a medium of high refractive index into a material with a lower refractive index. If the angle of incidence is greater than the critical angle then the light will be reflected.

track The total distance from the object to the image.

transmittance The ratio of the incident radiant power to the output radiant power.

transverse Any direction orthogonal to the optical axis.

tristimulus values An observer can match practically any colour by varying the amounts of red, green and blue present in a test patch placed alongside the colour being matched. The amounts of light required for the three matching colours are the tristimulus values.

ultraviolet (UV) Electromagnetic radiation with wavelengths below the lower-wavelength photopic limit and above the X-ray region. The optical worker usually considers the UV to lie between about 100 and 400 nm.

V-coating A very narrow band extension of the multilayer type antireflection coating, so named because of the V-shape of the transmission or reflectance curve. Coatings of this type are optimized for a specific wavelength and angle of incidence, so that they can reduce the reflectance to as low as 0.1 per cent over a narrow waveband. Although more expensive than a single layer, such coatings are particularly useful with narrow-band light sources such as lasers.

V-value See Abbe V-value.

veiling glare A manifestation of stray light without a recognizable image and having the effect of reducing contrast.

vertex The intersection point between the optical axis and the centre of the optical surface.

vignetting The partial obscuration of a ray bundle. It can cause a gradual reduction in lens system transmission as the off-axis angle increases.

virtual image This is formed when the object distance is less than the focal length of the lens. In this situation, rays leaving the lens appear to diverge from a point on the optical axis on the same side of the lens as the object. This point is the location of the virtual image.

vitreous humour In the human eye, the transparent gelatinous material between the lens and the retina.

watt (W) The unit of radiant power.

waveband Waveband is the spectral extent of a source, detector, filter or any optical component.

wavefront A surface of equal phase or equal optical distance from the origin of the light source. The rays emanating from a point source will be radial and therefore at right angles to the wavefront. In nearly all real situations light rays will be orthogonal to the wavefront, and only for some crystals will this not be true.

waveguide Any optical structure that will confine and support multiple reflections or the continuous propagation of light. This may take the form of a fibre optic cable, step index where multiple total internal reflections guide the light, or multimode where the refractive index profile maintains discrete propagation modes.

wavenumber The reciprocal of the wavelength when measured in centimetres.

wide-angle lens A lens with an effective focal length smaller than the diameter of the image it forms.

Wien's displacement law This states that for blackbody radiation, the spectral position of the peak of the radiant power is inversely proportion to its absolute temperature. So, as the temperature increases, the radiant power rises and the spectral peak moves to a shorter wavelength.

Wollaston prism The Wollaston, Nicol and Glan Thompson prisms all separate the two orthogonal polarization states of light through the use of a birefringent material such as calcite ($CaCO_3$). The Wollaston prism works best in collimated light.

zone The field position corresponding to 0.7 of the full field height.

zone plate A lens that forms a real image as a consequence of diffraction at the boundaries of a series of concentric opaque zones.

zoom lens A lens with a variable focal length but with a focal plane fixed in space.

APPENDIX B

List of symbols

In some cases a symbol can have more than one meaning but the correct choice should be obvious.

A_d	detector area
A_r	Airy disc radius
A_s	source area
B	blur spot diameter
B_w	Bandwidth
c	speed of light (3×10^8 m/s)
cv	Curvature of a spherical surface ($1/R$)
D	lens diameter
D_p	Optical power in dioptres
d	grating pitch
d_1	far depth of focus
d_2	near depth of focus
$\mathrm{d}u$	Small change in object distance
$\mathrm{d}v$	Small change in image distance
E_c	lx s or talbots m^{-2}
E_v	illuminance
f	focal length
f_b	Bragg cell frequency
f_c	spatial frequency (lp mm^{-1})
f_f	focal length of field lens
f_l	laser frequency
f_o	focal length of objective lens
$f\#$	f-number
h	Planck's constant (6.626196×10^{-34} J s)
h_1	object height
h_2	image height
I_f	forward current
I_v	luminous intensity
I_0	luminous intensity normal to emitting surface
I_θ	luminous intensity at an angle θ
i	angle of incidence
k	Boltzmann constant ($1.380\,622 \times 10^{-23}$ J K^{-1})
L	coherence length
L_v	luminance or brightness
M_v	luminous exitance
m	magnification
NA	numerical aperture
N	total number of grooves in a diffraction grating

n	refractive index
P_1	initial optical power
P_2	optical power after absorption
p	$= (y^2 + z^2)^{1/2}$
R	radius of curvature
R	total reflectivity
R_p	reflectivity for the p polarization state
R_s	reflectivity for the s polarization state
R_1	front radius of curvature of a lens
R_2	back radius of curvature of a lens
r	radius
S_1	front principal plane position, measured from the vertex
S_2	back principal plane position, measured from the vertex
T	absolute temperature
t	lens or glasss thickness
t_c	centre thickness of a lens
t_e	edge thickness of a lens
u	object conjugate distance
V_λ	photopic response
v	image conjugate distance
W	full width of a diffraction grating.
w	laser beam waist diameter ($1/e^2$ points)
X	sag of a spherical surface
x	distance along an optical axis
$x_\lambda, y_\lambda, z_\lambda$	colour-matching functions
y	tangential distance
y	normalized spatial frequency
z	saggital distance
α	absorption coefficient
$\delta\lambda$	wavelength difference
ν	frequency
$\delta\nu$	frequency bandwidth
θ	ray angle
θ_c	critical angle
θ_i	angle of incidence
θ_n	diffraction angle
θ_r	angle of refraction or reflection
λ	wavelength
σ	Stefan–Boltzmann constant ($5.669\,61 \times 10^{-8}\,\mathrm{W\,m^{-2}\,K^{-4}}$)
σ_r	marginal ray angle
Φ_e	radiant flux
Φ_v	luminous flux
ϕ	lens diameter
Ω	solid angle in steradians

APPENDIX C

Mathematical formulae

Basic optics

Snell's law

general case

$$n_1 \sin i = n_2 \sin r$$

from air to glass

$$\sin i = n_2 \sin r$$

optical thickness t_1 of a glass block of thickness t

$$t_1 = t/n$$

focal length of a thin lens

$$1/f = (n-1)(1/R_1 - 1/R_2)$$

for one plano surface

$$1/f = (n-1)/R$$

focal length (f_r) of a lens combination

$$1/f_r = f_1 f_2/(f_1+f_2-d)$$

focal length of a thick lens

$$1/f = (n-1)[1/R_1 - 1/R_2 + t(n-1)/nR_1R_2]$$

Note: use the thin-lens equation if $t \ll$ diameter (i.e. $>1:5$)

focal length from conjugate distances

$$1/v + 1/u = 1/f$$

conjugate distances at a fixed magnification

$$v = f(m+1)$$

$$u = (m+1)f/m$$

magnification

$$m = v/u = h_2/h_1$$

small longitudinal changes in image–object distance
(longitudinal magnification)

$$m^2\,du = dv$$

principle plane positions for a thick lens
(measured from the lens vertex)

$$S_1 = -f(n-1)t/R_2 n$$

$$S_2 = -f(n-1)t/R_1 n$$

f-number

$$f\# = f/\phi$$

marginal ray angle σ_r

$$\tan\sigma_r = \phi/(2f) \qquad \text{or} \qquad 1/2f\#$$

numerical aperture
general

$$NA = m\phi/2f(m+1)$$

$$NA = n\sin\sigma_r$$

for one infinite conjugate or when $m/(m+1) \approx 1$

$$NA = \phi/2f \qquad \text{or} \qquad 1/2f\#$$

lens power

$$\text{power(dioptres)} = 1/\text{focal length (in metres)}$$

intensity distribution of a far field diffraction pattern due to a circular aperture

$$I_{(z)} = I_0[2J_1(z)/z]^2$$

$$I_z = I_0[\sin cz]^2$$

where $J_1(z)$ is a Bessel function of the first kind, zero-order, and $z = \pi D\sin\theta/\lambda$. In this case D is the aperture diameter and θ is the angular radius from the centre of the interference pattern.

Airy radius

$$A_r = 1.22\,\lambda f\#\ (\mu m)$$

or

$$A_r = 1.22\,\lambda/(2\times NA)\ (\mu m)$$

modulation transfer function

$$MTF = (2/\pi)[\cos^{-1}(y) - y(1-y^2)^{1/2}]$$

MTF and CTF

$$MTF(v) = (\pi/4)[C(v) + C(3v)/3 - C(5v)/5 + C(7v)/7\ldots]$$

$$CTF(v) = (4/\pi)[M(v) - M(3v)/3 + M(5v)/5 - M(7v)/7 + M(9v)/9]$$

depth of focus

aberration-free case

$$d_1 = vB/(D-B) \qquad d_2 = vB/(D+B)$$

geometric case

$$\mathrm{DOF} = \pm\, Bf\#$$

(where $B = Z/\pi f_c$, $Z = 2.2$ for 50 per cent MTF)

hyperfocal distance for a projection lens

$$H = f + 1000f/f\#$$

near focal point (approx.)

$$d_2 = Hv/(H+v)$$

far focal point (approx.)

$$d_1 = Hv/(H-v)$$

sagitta of a spherical surface

$$X = R - (\mathrm{R}^2 - \mathrm{r}^2)^{1/2}$$

$$\mathrm{CV} = 2X/(r^2 + X^2)$$

spherometer formula

$$R = [1000(n-1)]/D_p$$

sag of an aspheric surface

$$X = \frac{cv \cdot p^2}{1+[1-(1+k)cv^2p^2]^{1/2}} + a_4p^4 + a_6p^6 + a_8p^8$$

The conic constant is defined as:

$$k = -\varepsilon^2$$

The conic surface is

1. An ellipse if $\varepsilon<1$.
2. A parabola if $\varepsilon=1$.
3. A hyperbola if $\varepsilon>1$.
4. A spheric if $\varepsilon=0$.

parabola (vertex at the origin)

$$Y^2 = 4fX$$

Increase in focal length of a parabola: F represents the distance from an off-axis point to the paraxial focal point:

$$F = f + x = f + Y^2/4f$$

ellipse

an ellipse with centre at (x_0, y_0) and major axis parallel to the x-axis is defined as

$$(x - x_0)^2/a^2 + (y - y_0)^2/b^2 = 1$$

length of major axis $= 2a$
length of minor axis $= 2b$

distance (c) from centre to focus (F or F') is given by

$$c^2 = a^2 - b^2$$

eccentricity is given by

$$\varepsilon^2 = c^2/a^2 = (a^2 - b^2)/a^2$$

Reflectivity and coatings

Reflection at a plane surface

$$R = (1-n_1/n_2)^2/(1+n_1/n_2)^2$$

for an air–glass boundary this reduces to

$$R = (n-1/n+1)^2$$

the s and p polarized reflections are given by

$$R_p=[\tan(\theta_1-\theta_2)/\tan(\theta_1+\theta_2)]^2$$

$$R_s=[\sin(\theta_1-\theta_2)/\sin(\theta_1+\theta_2)]^2$$

where θ_1 is the angle of incidence and θ_2 is the angle of refraction

Brewster angle

$$\theta_b = \arctan(n_2/n_1)$$

critical angle

$$\theta_c = \arcsin(n_2/n_1)$$

centre wavelength shift of a coating versus angle of incidence

$$\lambda_\theta=\lambda[n^2 - n_0 \sin^2\theta]^{1/2}/n$$

where λ is the nominal peak of the filter and λ_θ is the peak at an angle θ

Dispersion and gratings

Abbe V-value

$$V_d = \frac{n_d - 1}{n_F - n_C}$$

diffraction from a grating

normal incidence

$$d \sin\theta = m\lambda$$

in this case, m is the diffraction order $m = 1, 2, 3$, etc.

if the angle of incident light is finite then

$$d\sin(\theta_n - \theta_i) = m\lambda$$

resolving power of a grating

$$\lambda/\delta\lambda = mN \qquad \text{where } N = W/d$$

absorption

$$P_2 = P_1 e^{-\alpha x}$$

Interference

constructive interference in an etalon

$$m\lambda = 2nd\cos\theta$$

where m = an integer referred to as the order
d = the thickness of the cavity
θ = the angle of incidence to the cavity

Photometry and radiometry

ASA and DIN ratings

$$\text{ASA rating} = 5/E_c$$

$$\text{DIN rating} = (10\log E_c) - 5$$

luminous intensity

$$I_v = \Phi_v/\Omega$$

or

$$I_v = E_v d^2$$

for a Lambertian emitter

$$I_\theta = I_0\cos\theta$$

solid angle

$$\Omega \approx A_d/d^2$$

for a circularly symmetric area

$$\Omega = 2\pi\,[1 - \cos\theta]$$

illuminance

$$E_v = \Phi_v/A_d$$

or

$$E_v = \Phi_v/4\pi d^2$$

cosine fourth law

$$\Phi_v = L_v A_s A_d \cos^4\theta/d^2$$

image illuminance

$$E_v = L_v A_l \cos^4(\theta)/f^2$$

for a circular lens

$$E_v = L_v \pi \cos^4 \theta / f\#^2$$

luminous exitance

$$M_v = \int_0^{\pi/2} 2\pi L_v \cos\theta \sin\theta \, d\theta = \pi L_v$$

luminance

$$L_v = I_v / A_s$$

for a point source

$$L_v = \Phi_v / A_s 4\pi$$

spectral luminous efficacy

$$K(\lambda) = \Phi_{v\lambda} / \Phi_{e\lambda} \ (\text{lm W}^{-1})$$

spectral luminous efficiency

$$V(\lambda) = 683K(\lambda)$$

luminous efficacy

$$K(\lambda) = 683\int_0^{\infty} V(\lambda)\Phi_{e\lambda} \, d\lambda \Big/ \int_0^{\infty} \Phi_{e\lambda} \, d\lambda$$

For a blackbody

$$\mathrm{K}(\lambda) = 683\int_0^{\infty} V(\lambda)\, M(\lambda) \, d\lambda / \sigma T^4$$

Colour

CIE colour coordinates (1931)

$$x = X/(X+Y+Z)$$
$$y = Y/(X+Y+Z)$$
$$z = Z/(X+Y+Z)$$

Non-monochromatic case

$$X = \int_{380}^{780} \phi(\lambda) \times x(\lambda) \, d\lambda$$
$$Y = \int_{380}^{780} \phi(\lambda) \times y(\lambda) \, d\lambda$$
$$Z = \int_{380}^{780} \phi(\lambda) \times z(\lambda) \, d\lambda$$

CIE colour coordinates (1976)

$$u' = 4X/(X+15Y+3Z)$$
$$v' = 9Y/(X+15Y+3Z)$$

CIE colour coordinates (1960)

$$u=u' \qquad \text{and} \qquad 1.5v=v'$$

CIELUV

CIELUV space is based on the 1976 u' v' chromaticity diagram and is generated using the quantities L^*, u^* and v^*, calculated from:

$$L^*=116(Y/Y_n)^{1/3}-16$$

$$u^*=13L^*(u'-u'_n)$$

$$v^*=13L^*(v'-v'_n)$$

Y, u' and v' are for the colour being measured, and Y_n, u'_n and v'_n are for an appropriate white reference.

If Y/Y_n is less than 0.008 856 then L^* is calculated from $903.3Y/Y_n$ instead. If two colours differ in L^*, u^* and v^* by dL^*, du^* and dv^*, then colour difference dE^* between them is given by:

$$dE^*_{uv} = [(dL^*)^2+(du^*)^2+(dv^*)^2]^{1/2}$$

L^*, u^* and v^* can also be used to calculate values for the quantities which influence colour perception, brightness, hue and saturation.

lightness$_{uv}$

$$L^*=116(Y/Y_n)^{1/3} - 16\,(Y/Y_n > 0.008\,856)$$

A change in L^* represents more or less the same change in lightness for any colour.

saturation$_{uv}$

$$S_{uv} = 13[(u'-u'_n)^2+(v'-v'_n)^2]^{1/2}$$

S_{uv} is based on the CIE 1976 UV chromaticity diagram, so it is a good match to saturation.

chroma$_{uv}$

$$C_{uv}=S_{uv}L^*$$

The combination of saturation and lightness allows for the fact that chrominance decreases as the luminance falls, but the use of the tristimulus values means that chroma is not affected by changes in the illuminance.

hue angle$_{uv}$

$$h_{uv} = \tan^{-1}(v^*/u^*)$$

Here the use of u and v mean that the hue angle corresponds closely to perceived hue.

hue difference$_{uv}$

$$dH^*_{uv}=[(dE^*_{uv})^2 -(dL^*)^2-(C_{uv})^2]^{1/2}$$

CIELAB

The CIELAB space is similar to CIELUV but has no saturation and is not based on a chromaticity diagram. The quantities L^*, a^* and b^* are given by

$L^* = 116(Y/Y_n)^{1/3} - 16$ where $Y/Y_n > 0.008\,856$

$a^* = 500[(X/X_n)^{1/3} - (Y/Y_n)^{1/3}]$ where $X/X_n > 0.008\,856$, $Y/Y_n > 0.008\,856$

$b^* = 200[(Y/Y_n)^{1/3} - (Z/Z_n)^{1/3}]$ where $Z/Z_n > 0.008\,856$

As before, the subscript n refers to a reference white. Colour differences are given by

$$\mathrm{d}E^*_{ab} = [(\mathrm{d}L^*)^2 + (\mathrm{d}a^*)^2 + (\mathrm{d}b^*)^2]^{1/2}$$

chroma$_{ab}$

$$C_{ab} = (a^{*2} + b^{*2})^{1/2}$$

hue angle$_{ab}$

$$h_{ab} = \tan^{-1}(b^*/a^*)$$

hue difference$_{ab}$

$$\mathrm{d}H^*_{ab} = [(\mathrm{d}E^*_{ab})^2 - (\mathrm{d}L^*)^2 - (C_{ab})^2]^{1/2}$$

Detectors

noise equivalent power

$$\mathrm{NEP} = \text{noise current (A)/photodiode sensitivity (A W}^{-1})$$

specific detectivity

$$D^* = (\sqrt{A_\mathrm{d}} \times \sqrt{B_\mathrm{w}})/\mathrm{NEP}$$

Fibres

attenuation

$$\text{attenuation} = 10\log(\text{input power/output power})$$

optical fibre numerical aperture

$$\mathrm{NA} = n_\mathrm{air} \sin(\sigma_\mathrm{r})$$

$$\mathrm{NA} = (n_\mathrm{core}^2 - n_\mathrm{cladding}^2)^{1/2}$$

Light sources

Planck's law for the energy of a quantum

$$E = hc/\lambda$$

coherence length of a laser

$$L = c/\delta\nu$$

natural divergence of a laser

$$\text{divergence} = 1.22\lambda/w$$

laser beam radius of curvature $R(z)$

$$R(z) = z[\,1 + (\pi r_0^2/\lambda z)^2\,]$$

where z is the distance travelled from the waist and $r_0 = w/2$ for this case

laser beam diameter ($1/e^2$ point) at distance z

$$r(z) = r_0\,[\,1 + (\lambda z/\pi r_0^2)^2\,]^{1/2}$$

where z is the distance travelled from the waist and $r_0 = w/2$ for this case

radiant exitance of a blackbody

$$M(\lambda) = 2\pi c^2 h/\lambda^5(e^{hc/\lambda kT} - 1)\ \text{W m}^{-2}\,\text{m}^{-1}$$

spectral radiance of a blackbody

$$L(\lambda) = 2c^2 h/\lambda^5(e^{hc/\lambda kT} - 1)\ \text{W m}^{-2}\,\text{sr}^{-1}\,\text{m}^{-1}$$

total luminance of a blackbody

$$L_v = 2h\nu^3/c^2(e^{h\nu/kT} - 1)\ \text{W m}^{-2}\,\text{sr}^{-1}\,\text{Hz}^{-1}$$

If $hc \ll$ kT

$$L_v = c_1 c_2 T \nu^2 \lambda^{-4}$$

If $hc \gg$ kT

$$L_v = c_1 \lambda^{-5} e^{-hc/\lambda kT}$$

total exitance of a blackbody

$$M_{\text{total}} = \sigma T^4$$

Further reading

Born, M. and Wolf, E. (1970) *Principles of Optics,* 4th Edition, Pergamon Press, Oxford.

Chaimowicz, J. C. A. (1992) *Opto-Electronics: an introduction,* Butterworth-Heinemann, Oxford.

Chamberlin, G. J. and Chamberlin, D. G. (1980) *Colour: its measurement, computation and application,* Heydon & Sons Ltd., London.

CIE Publication 18.2 (1983) *The Basis of Physical Photometry*, 2nd edition.

CIE Publication 15.2 (1986) *Colorimetry*, 2nd edition.

CIE Publication 17.4 (1987) *International Lighting Vocabulary*, 4th edition.

Dereniak, E. L. and Crowe, D. G. (1984) *Optical Radiation Detectors,* John Wiley and Sons, Chichester.

Heavens, O. S. and Ditchburn, R. W. (1991) *Insight into Optics*, John Wiley and Sons, Chichester.

Hecht, E. (1975) *Theory and Problems of Optics,* McGraw-Hill, New York.

Hecht, J. (1992). *The Laser Guidebook,* 2nd edition, McGraw-Hill, New York.

Hecht, E. and Zajac, A. (1979) *Optics,* 4th edition, Addison Wesley.

Kingslake, R. (1978) *Lens Design Fundamentals,* Academic Press, New York.

Kingslake, R. (1983) *Optical Systems Design,* Academic Press, New York.

Longhurst, R. S. (1967) *Geometrical and Physical Optics,* 2nd edition, Longmans, Green and Co. Ltd., London.

Malacara, D. (ed.) (1978) *Optical Shop Testing,* John Wiley and Sons, Chichester.

Shannon, R. and Wyant, J. (1980) *Applied Optics and Optical Engineering,* vol. 8, Academic Press, New York.

Smith, W. J. (1990) *Modern Optical Engineering,* 2nd edition, McGraw-Hill, New York.

Sobel, M. (1989) *Light*, University of Chicago Press, Chicago.

Welford, W. (1988) *Optics,* 3rd edition, Oxford University Press, Oxford.

Welford, W. and Winston, R. (1989) *High Collection Non-imaging Optics,* Academic Press, London.

Index

Abbe V value, 66
Abbe sine condition (*see* sine condition)
Aberration, 25–33
(*see also* astigmatism, coma, distortion, field curvature, lateral colour, longitudinal colour, spherical)
Absorption, 81
Achromatic, 38
Adaptation (*see* light and dark adaption)
Afocal, 42
Airy disc, 20, 36
Ambient light, 134
Amici prism, 69
Anamorphic, 183
Angles of incidence, 55–56, 64, 74, 78, 85
refraction, 2–4
reflection, 87
Anisotropic crystal, 34
Anomalous dispersion, 66
Antireflection coating, 57
Apochromatic, 41
Apostilb, 113
ASA rating, 107
Aspheric:
lens, 49
surface, 27
Astigmatism, 28
circle of least confusion, 26
sources, 29, 183
Athermalization, 40
Attenuation, 138

Back focal length, 14
Baffles, 134
Band pass filter, 62
Barrel distortion, 29
Beamsplitter:
cube, 71
Beam dump, 210
Beam expander, 8
Beam waist, 165
calculation of, 166
Bend radius, 140
Biconcave, 8
Biconvex, 8
Birefringence, 72
Blackbody radiation, 185–189
Blur circle, 23, 26
Boltzmann's constant, 187
Borosilicate, 123
Bragg cell, 155
Brewster:
angle, 56
window, 166
Brightness, 113
Buyers' guide, 100–103

Calcite crystal, 72
Candela, 107
Cardinal points (*see* principal planes)
Cassegrain, 53
Catadioptric, 52
Catoptric, 53
CCD camera, 132–133
CCTV lens, 104
Centring errors, 198–200
Chief ray, 15
Chromatic aberration (*see* lateral colour and longitudinal chromatic)
Chromaticity:
coordinates, 118–119
diagrams, 119
1931, 120
1960, 119, 189
1976, 121
Circle of least confusion, 26
(*see also* blur circle)
Circularly polarized light, 34
Coatings:
antireflection, 57, 202–203
broadband, 58
cold, 63
dichroic, 60
full width at half maximum, 59–60
GRIN, 64
hot, 63
interference, 59
multilayer, 59
narrow band, 58
single layer, 57
specification of, 64, 203
V-coat, 58
Coherence length, 163
bundle, 153
radiation, 161
Cold mirror, 63
Colour, 118
measurement, 118
Colour coordinates (*see* chromaticity coordinates)
Colour differences, 121

Colour matching functions, 118, 137
Colour temperature, 188–189
Coma, 27
Concentricity, 198
Condenser, 50
Cones, 114
Conic constant, 50, 27
Conjugates, 9, 11
Contrast, 21
Contrast transfer function, 21
Core, 139, 141
Correlated colour temperature, 188–189
Cosine laws, 109–110
Critical angle, 56
Cube beamsplitter, 71
Custom design, 104–105
Cylindrical lens, 182, 183

Dark adaption, 114
Dark noise, 129
Decibel, 138
Depth of field (*see* depth of focus)
Depth of focus, 23–24
  aberration free, 23
  geometric, 24
Detectivity, 129–130
Detectors, 121
  infrared, 131–132
  ultraviolet, 122–126
  visible, 126–131
Dichroic filters, 60
Diffraction:
  Fraunhofer, 20
  Fresnel, 20
Diffraction grating:
  amplitude grating, 77
  blazed, 77
  equation, 74
  manufacture, 78
  order, 74
  phase, 78
  reflecting, 76
  resolving power, 75
  transmitting, 76
Diffraction limited performance (*see* Airey disc)
Diffractive lens:
  zone plate, 78–82
  holographic, 82
DIN index, 107
Dioptre, 19
  gauge, 198
Dispersion, 2, 65
  anomalous, 67
  modal, 141
Distortion, 29
  TV and film, 30
Divergence, 165
Doublet, 38, 40
Double refraction (*see* polarization)
Dove prism, 71

Eccentricity, 27, 50, 95
Edge filter, 60
Effective focal length, 14
Electro-brightening, 98
Energy, 107
Entrance pupil, 17
Etalon, 85
Etendue (*see* Lagrange invariant)
Exit pupil, 17
Exitance, 112
Extraordinary ray, 34
Eye, 114
Eyepiece, 48
Eye relief, 49

F number (f#), 18
Fabry–Pérot:
  etalon, 85
Ferrule, 147
Fibre bundle:
  coherent, 153
  incoherent, 151
Fibre optic, 138
  applications, 152–157
  bend radius, 140
  connectors, 145–149
  graded index (*see* multimode)
  launching, 151
  mode stripping, 157
  monomode, 144
  mounting, 208
  multimode, 142
  plastic, 144–145
  preparation, 150–151
  sensors, 153–157
    intensity, 154
    phase, 155–157
    polarization, 155
    wavelength, 154
  step index, 141
Field angle (*see* field height)
Field curvature, 31
Field height, 16
Field lens, 49, 135
Field of view, 135
Fluorescence, 184
Flux density, 108
Focal length:
  back focal length, 14
  effective, 14
  front focal length, 14
Foot candle, 112
Footlambert, 113
Form error, 189
Fovea, 114
Fraunhofer diffraction, 20
Frequency doubling, 174
Fresnel:
  diffraction, 20
  lens, 51
  reflection, 55
Front focal length, 14

Full width half maximum, 59–60
Fused silica, 123

Gaussian beam:
  intensity profile, 20, 165
Gaussian optics, 11
Geometric optics (*see* paraxial optics)
Ghost image, 41
Glan Thompson prism, 72
Graded index fibre, 142
GRIN lens, 143, 149

Heat absorbing filter, 63
Helium-Neon laser, 172
Hot mirror, 63
Hue, 118 (*see also* chromaticity)

Illuminance, 109
  and lenses, 110
  conversion factors, 112
  measurement, 111
Illuminant A, 116
Images:
  real, 12
  virtual, 12
Incandescent lamp (*see* lamps)
Incoherent radiation, 163
Infrared, 122
Insertion loss, 150
Integrating sphere, 107
Intensity (*see* luminous intensity)
Interference, 161
  constructive, 162
  destructive, 162
Interference filter, 59–60
Interferometer:
  Fizeau type, 93, 195
  Twyman–Green, 164, 195
Inverse square law, 109
Irradiance (*see* illuminance)
Isotropic crystal, 34

Kinematic mount, 147

Lagrange invariant, 113
Lambert, 113
Lambert's law, 108
Lambertian emitter, 113
Lamps:
  design guide, 190–191
  discharge, 189–190
  filament, 185
  fluorescent, 184
  long arc, 191
  short arc, 190–191
  spectral, 192
  tungsten-halogen, 186
Lasers, 158–174

Lasers – *continued*
  beam waist calculation, 166
  coherence, 161
  expanding, 168
  focus, 167
  mode, 165
  polarization, 166
  safety, 158–159
  specification of, 168–169
Laser types, 169–174
  CO_2, 93, 98, 173–174
  diode, 180–184
    threshold current, 181
    divergence, 182
  Helium-Cadmium, 171
  Helium-Neon, 172
  infrared, 170
  Nd-YAG, 172
  ultraviolet, 170
  visible, 170
Lateral colour, 32
LED, 175–180
  edge emitting, 180
  emission angle, 177
  forward voltage, 177
Lens:
  aspheric, 49
  doublet, 38
    guidelines for use, 40
  drawing, 193–203
  Fresnel, 51
  mirror, 52
  mounting, 204–206
  singlet, 35
    guidelines for use, 38
  surface:
    defects, 201
    texture, 201–202
    treatment, 202–203
  telephoto, 42
  triplet, 40
    guidelines for use, 41
  wide angle, 43
Lens nomenclature, 17
Light adaptation, 114
Line spread function, 23
Linewidth, 163
Longitudinal chromatic, 32
Lumen, 107
  lighting design (LDL), 112, 184
  screen, 111
Luminance, 113
  conversion factors, 114, 115
Luminous efficacy, 117
Luminous efficiency, 117
Luminous intensity, 107, 176
Lux, 108

Magnification:
  linear, 10
  longitudinal, 11

Mangin mirror, 53
Marginal ray, 16
  angle, 18
Meniscus lens:
  concave, 9
  convex, 9
Mercury arc, 191
Meridional (*see* tangential)
Mesopic, 116
Metal halide arc, 191
Microscope, 45
  compound, 45
  eyepiece, 48
  objectives, 48
  parfocal, 46
  RMS thread, 47
Milliphot, 112
Mirror:
  basics, 83
  commercial, 98
  concentrators, 96
  condenser, 88
  ellipse, 94
    equations, 95
  equation, 87
  free form, 96
  materials, 97, 203
  parabolas, 89, 91
    off axis, 92
  scientific, 97
  spherical, 86, 91
Modal dispersion, 141
Modulation transfer function, 21
  diffraction limited, 22
  geometric, 23
Monochromatic (*see* lasers)
Monochromator, 67, 76
Monomode fibre optic, 144
Multilayer coating, 59–60

Newtonian telescope, 53
Newton's rings, 195–196
Nicol prism, 72
Nit, 114
Nodal points, 12
Nodal slide, 16
Noise:
  Johnson, 129
  shot, 129
Noise equivalent power, 129
Numerical aperture (NA), 18

Objective lens, 48
Optical nerve, 114
Optical path, 5
Optical thickness, 3
Ordinary ray, 34

Paraxial optics, 6
Parfocal, 46
Pellicle beamsplitter, 71
Penta prism, 71
Petzval surface, 32
  (*see also* field curvature)
Phot, 112
Photodiode:
  avalanche, 131
  PIN, 123, 128
  p–n, 127
  vacuum, 123
    cylindrical, 124
    planar, 124
Photoelectric effect, 126
Photoemission, 123 h+2
Photometer, 111
Photometry, 106
Photomultiplier, 125
Photopic vision, 114, 136
Pitch of GRIN lens, 149
Planck's law, 160
Planck's constant, 160, 187
Plank's radiation law, 187
Plane polarized, 33
Planoconcave, 9
Planoconvex, 8
Plastic injection moulding, 98
Point source, 108
Polarization, 33
  (*see also* polarizing prisms)
Polarizing prisms, 73
Polish numbers, 202
Porro prism, 71
Power, 107
  lens:
    dioptres, 19
Principal planes, 12
  positions, 14–15
Prisms:
  Amici, 69
  cube, 71
  dove, 71
  mounting, 206–209
  penta, 71
  porro, 71
  rhomboid, 69
  right angle, 68
  Wollaston, 72
Projection lens, 88
Pulse spreading, 140
Pupil:
  entrance, 17
  exit, 17
Pyroelectric detector, 131

Quarter wave plate, 34
Quartz, 129

R_a, 84, 201
Radiance (*see* luminance and brightness)
Radiation constants:
  1st and 2nd, 187

Radiometry:
  measurement, 134
Ray nomenclature, 16
Rayleigh's criteria, 47
Rayleigh–Jeans approximation, 188
Rayleigh limit, 47
Reflectivity:
  Fresnel, 55
  p polarization, 55
  s polarization, 55
Reflecting telescopes:
  Cassegrainian, 90
  Newtonian, 90
Refraction, 2
Refractive index, 1
  GRIN profile, 149
Resolution, 19
Resolving power of a grating, 75
Responsivity, 129
Retardation plate, 33–34
Retina, 114
Rhomboid prism, 69
Right-angle prism, 68
Rods, 114
Roughness numbers, 202
Royal Microscopical Society (RMS) thread, 47

Sag, 27
Sagittal, 17
Sapphire, 123
Schmidt corrector (plate), 92
Scotopic, 114, 116
Scratch and dig, 84, 201
Secondary colour, 67
Semiconductor laser (*see* laser diode)
Sign conventions, 7
Silicon:
  detector, 126
    spectral response, 127
Sine condition, 6
Skew ray, 17
Snell's law, 3
Solid angle, 107
Spatial frequency, 19, 24
Speckle, 164
Spectral radiant exitance, 187
Spectral luminous efficacy, 117
Spherical aberration, 25
  order of, 27
  overcorrected, 26
  transverse, 26
  undercorrected, 26
Spherometer, 19, 196–197
Spot diagram, 23
Stefan–Boltzmann law, 117, 188
  constant, 187
Step index fibre, 141
Steradian, 107
Stilb (sb), 115
Stimulated emission, 160
Stop, 17
Stray light, 78, 134

Talbot, 107
Tangential, 17, 28–29
Telephoto lens, 42
  inverse, 42–43
Telescope:
  Cassegrainian, 53
  Gregorian, 95
  Newtonian, 53
TEM_{00}, 165
Test plate, 194
Thick lens:
  equation, 10
  nodal points, 12
  principal plane positions, 14
Thin lens:
  equation, 9, 36
Total internal reflection, 56, 68
Tristimulus values, 119

Ultraviolet, 122
  UVA, 123
  UVB, 123
  UVC, 123

V-coating (*see* coatings)
Virtual image, 7
V value (*see* Abbe V value)

Watt, 107
Wavefront:
  addition/subtraction, 162
Wide angle lens, 43
Wien's:
  law, 188
  displacement law, 188
Wollaston prism, 72

Xenon arc, 191

Young's two slit experiment, 161

Zone, 27
Zone plate, 79–82
Zoom lens, 104